Lib

RICAN GROUND

Pl

5ℓ

0 3

GREENWICH LIBRARIES

3 8028 01564988 9

WORLD TRADE CENTER AND ENVIRONS

Destroyed

Heavily Damaged/Later Demolished

Heavily Damaged/Repairable

PATH Trains

NYC Subways

Slurry Wall

0   1/8   1/4

Miles

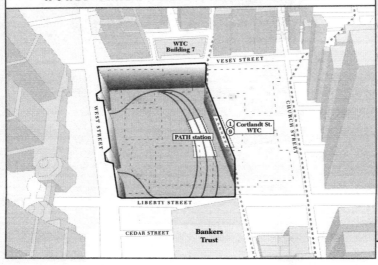

WORLD TRADE CENTER FOUNDATION HOLE

Public School 89

WF Building

North Cove Marina

WTC Building 7

VESEY STREET

WEST STREET

CHURCH STREET

① ⑨ Cortlandt St. WTC

PATH station

LIBERTY STREET

CEDAR STREET

Bankers Trust

Hudson River

NORTH END AVENUE

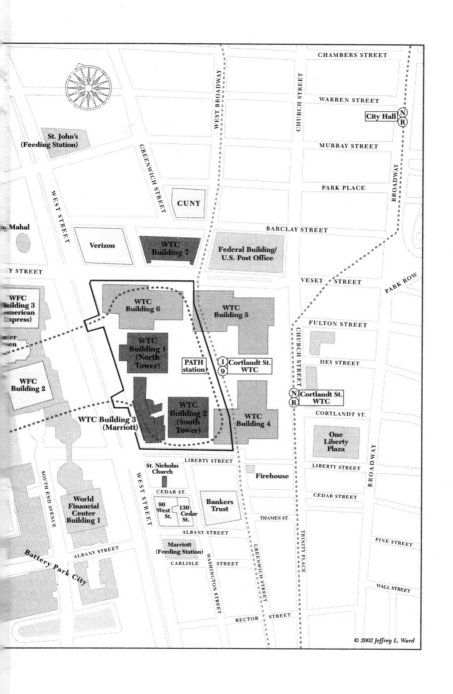

ALSO BY WILLIAM LANGEWIESCHE

Cutting for Sign
Sahara Unveiled
Inside the Sky

AMERICAN GROUND

AMERICAN GROUND

Unbuilding the World Trade Center

WILLIAM LANGEWIESCHE

Scribner

First published in the United States by Farrar, Straus and Giroux, 2002
First published in Great Britain by Scribner, 2003
An imprint of Simon & Schuster UK Ltd
A Viacom Company

Copyright © William Langewiesche, 2002

This book is copyright under the Berne Convention
No reproduction without permission
® and © 1997 Simon & Schuster Inc.
All rights reserved

Scribner and design are trademarks of Macmillan
Library Reference USA, Inc., used under licence by
Simon & Schuster, the publisher of this work.

Grateful acknowledgement is made to *The Atlantic Monthly*,
where this book originated as a three-part series.

The right of William Langewiesche to be identified as author of
this work has been asserted by him in accordance with
sections 77 and 78 of the Copyright, Designs and Patents Act, 1988.

1 3 5 7 9 10 8 6 4 2

Simon & Schuster UK Ltd
Africa House
64-78 Kingsway
London WC2B 6AH

Simon & Schuster Australia
Sydney

www.simonsays.co.uk

Designed by Jonathan D. Lippincott
Map designed by Jeffrey L. Ward

A CIP catalogue record for this book is available from the British Library

ISBN 0-7432-3969-5

Printed and bound in Great Britain by
Mackays of Chatham PLC

To Matthew and Anna

GREENWICH LIBRARIES	LOC
INV BROWNS 28.3.03 £8.99	
ACC NO 38028 01564 9889	
CAT 974.71	

CONTENTS

THE INNER WORLD

When the Twin Towers collapsed, on the warm, bright morning of September 11, 2001, they made a sound heard variously around New York as a roar, a growl, or distant thunder. The South Tower was the first to go. At 9:59 its upper floors tilted briefly before dropping, disintegrating, and driving the building straight down to the ground. The fall lasted ten seconds, as did the sound. Many people died, but mercifully fast. Twenty-nine minutes later the North Tower collapsed just as quickly, and with much the same result. Somehow a few people survived. For an instant, each tower left its imprint in the air, a phantom of pulverized concrete marking a place that then became a memory. Prefabricated sections of the external steel columns tumbled down onto lesser buildings, piling onto terraces and rooftops, punching through parking structures, offices, and stores, inducing secondary collapses and igniting fires. The most catastrophic effects were eerily selective: with the exception of Saint Nicholas, a tiny Greek Orthodox church that dissolved in the rain of steel, the only buildings completely wrecked were those that carried the World Trade Center label. There were seven in all, and ultimately none of them endured. Not even the so-called World Trade Center Seven, a relatively new forty-seven-floor tower that stood independently across the street from the complex, was able to escape the fate associated with its name. Though it did not seem seriously wounded at first, it burned persistently throughout the day, and that evening became the first steel-frame high-rise in history to fall solely because of fire.

There was wider damage, of course, and on the scale of ordinary disasters it was heavy. For thirty years the Twin Towers had stood above the streets as all tall buildings do, as a bomb of sorts, a repository for the prodigious energy originally required to raise so much weight so high. Now, in a single morning, in twin ten-second pulses, the towers released that energy back into the city. Massive steel beams flew through the neighborhood like gargantuan spears, penetrating subway lines and underground passages to depths of thirty feet, crushing them, rupturing water mains and gas lines, and stabbing high into the sides of nearby office towers, where they lodged. The phone system, the fiber-optic network, and the electric power grid were knocked out. Ambulances, cars, and fire trucks were smashed flat by falling debris, and some were hammered five floors down from the street into the insane turmoil erupting inside the World Trade Center's immense "bathtub"—a ten-acre foundation hole, seventy feet deep, that was suffering unimaginable violence as it absorbed the brunt of each tower's collapse.

The energy released within that wild, inaccessible core lit fires that cooked the ruins for months afterward. Outside of each tower's footprint, and still within the foundation hole, it demolished most of the six-story subterranean structure—consisting largely of parking garages that were either pulverized or badly broken and left to hang. Deep underground it also destroyed part of the Port Authority Trans-Hudson (PATH) commuter line— a railroad from New Jersey that, having passed in a single-track tube through the watery muck of the river's bottom, emerged into the foundation hole and traveled to a station on the far side before looping back to a parallel tube and returning under the river to New Jersey. The PATH tubes were century-old cast-iron structures, probably brittle in places, and now at immediate risk of failure. If either of them broke catastrophically, the Hudson River would flood into the foundation hole, filling it at high tide to a level just five feet below the street and drowning unknown numbers of trapped survivors. Moreover, on the far side of the river a wall of water would flood the Jersey City station, and from there, via connecting rail links, would circle uncontrollably back into Manhattan, rush through the passages beneath Greenwich Village, and take out the West Side subways from the southern tip of the island nearly to Central Park. Vulnerability to sequential flooding was a known weakness of the PATH system, and it had been highlighted in a report circulated discreetly among government officials after the earlier World Trade Center attack, the parking-garage bombing of February 26, 1993. But maybe because such flooding was also something of an apocalyptic vision—and therefore somehow unreal—no defenses were erected against it. Of course now it was too late. And immediately as the Twin Towers collapsed, it became obvious that even in America apocalypses could come to pass.

On the surface the scene was just as rough. At the southwest corner of the World Trade Center complex the twenty-two-floor Marriott hotel was transformed into a raw, boxy thing three stories high. Just to the north, across West Street, a pedestrian bridge gave way, killing groups of firemen and office workers who had sheltered beneath it. The streets buckled under heaps of smoking steel. So much heavy debris fell across the access routes that rescue vehicles were rendered useless. Major fires ignited in all directions. Simultaneously, air-pressure waves shifted small cars and shattered windows for several blocks around, blowing powdery World Trade Center remains into apartments and across the chest-high partitions of corporate offices. The powder was made primarily of crushed concrete. The waves generated winds that pushed it through the streets in dense, choking clouds and lifted it to mix with smoke and darken the morning. Then all the white paperwork floated down on the city as if in mockery of the dead.

The suddenness of the transformation was difficult to accept. It had taken merely one brief morning, merely twenty seconds of collapse, and now all that remained standing of the Twin Towers were a few skeletal fragments of the lower walls, the vaguely gothic structures that reached like supplicating hands toward the sky. After the dust storms settled, people on the streets of Lower Manhattan were calm. They walked instead of running, talked without shouting, and tried to regain their sense of place and time. Hiroshima is said to have been similar in that detail. The site itself remained frightening because of the confusion of ruins and fire, as well as the possibility of further attacks or collapses. But a reversal soon occurred by which people began moving toward the disaster rather than away from it. The reaction was largely spontaneous, and it cut across the city's class lines as New Yorkers of all backgrounds tried to respond. A surprising number of stockbrokers, shopkeepers, artists, and others got involved. For the most part, however, it was the workers with hardhats, union cards, and claims to a manual trade who were able to get past the police checkpoints that had been established earlier that morning, after the first airplane struck. Few of these workers lived in Manhattan, though typically they were there that morning on jobs. They hailed from Staten Island, Brooklyn, the Bronx,

Queens, and the close-in suburbs of Long Island and New Jersey, and most had accents to prove it. From the start, therefore, the recovery site was what it remained: an outer-borough New York blue-collar scene—overwhelmingly Irish, Italian, and male, terribly unrepresentative by social measures, and yet authentic.

Arriving at the site over the first few hours, the volunteers joined with the firefighters and the police, who by then were shaking off their disbelief and struggling to take effective action. By afternoon thousands of people in these combined forces were searching through the ruins for survivors, attacking the debris by hand, forming bucket brigades, and climbing over the smoking pile that in some places rose fifty feet above the street. At 5:20 P.M. World Trade Center Seven collapsed tidily in place, damaging some adjacent buildings but killing no one. By dark the first clattering generators lit the scene, and an all-American outpouring of equipment and supplies began to arrive. The light stuff got there first: soda pop and bottled water, sandwiches, flashlights, bandages, gloves, blankets, respirators, and clothes. Indeed, there were so many donations so soon that the clutter became a problem, hindering the rescue effort, and a trucking operation was set up just to haul the excess away.

People who came to the site in those early days often had the same first sensation, of leaving the city and walking into a dream. Many also felt when they saw the extent of the destruction that they had stumbled into a war zone. "It's like something you'd see in the movies," people said. Probably so, but my own reaction was different when I first went in, soon after the attacks. After years of traveling through the back corners of the world, I had an unexpected sense not of the strangeness of this scene but of its familiarity. Wading through the debris on the streets, climbing through the newly torn landscapes, breathing in the mixture of smoke and dust, it was as if I had wandered again into the special havoc that failing societies tend to visit upon themselves. This time they had visited it upon us. The message seemed to be "Here's a sample of our political science." I was impressed by how faithfully the effects had been reproduced on the ground.

But you could never confuse New York with a back corner of the world, and the ruins did not actually look like a war zone either. There was sadness to the site, to be sure, and anger, but there was none of the emptiness—the ghostly quality of abandonment—that lurks in the aftermath of battle. In fact, quite the opposite quality materialized here: within hours of the collapse, as the rescuers rushed in and resources were marshaled, the disaster was smothered in an exuberant and distinctly American embrace. Despite the apocalyptic nature of the scene, the response was unhesitant and almost childishly optimistic: it was simply understood that you would find survivors, and then that you would find the dead, and that this would help their families to get on with their lives, and that your resources were unlimited, and that you would work night and day to clean up the mess, and that this would allow the world's greatest city to rebuild quickly, and maybe even to make itself into something better than before. From the first hours these assumptions were never far away.

For a few days the site was out of control. The bucket brigades were ineffectual, and barely scratched the surface of the ruins—not through lack of trying, God knows, but because of the overwhelming weight of the debris. In the end it probably didn't matter, because, as later became apparent, the dead did not die lingering deaths. At the time, however, this was neither known nor knowable. Very little was. Rumors swept the exhausted crowds of workers, and on multiple occasions caused dangerous stampedes away from the imagined reach of One Liberty Plaza, a perfectly sound building that was said to be falling. People were hurt in those panics. There were too many volunteers and too few heavy machines.

But then, rather quickly, a crude management structure was agreed upon, and most of the volunteers were eased out to the ruins' periphery, to be replaced at the core by a professional labor force that might loosely respond to direction—firemen and cops on overtime, structural and civil engineers, and up to 3,000 unionized construction workers.

The city government ran the show. The agency charged with managing the physical work was an unlikely one. It was the Department of Design and Construction (DDC), an obscure bureaucracy 1,300 strong whose normal responsibility was to oversee municipal construction contracts—for sidewalk and street repairs, jails, and the like—and whose offices were not even in Manhattan but in Queens. The DDC was given the lead for the simple reason that its two top officials, a man named Kenneth Holden and his lieutenant, Michael Burton, had emerged from the chaos of September 11 as the most effective of the responders. Now they found themselves running a billion-dollar operation with the focus of the nation upon them.

Nearly everyone at the site was well paid. The money for the effort came from federal emergency funds, and it flowed freely. But despite some cases of corruption and greed, money was not the main motivation here—at least not until almost the end. Throughout the winter and into the spring the workers rarely forgot the original act of aggression, or the fact that nearly 3,000 people had died there, including the friends and relatives of some who were toiling in the debris. They were reminded of this constantly, not only by the frequent discovery of human remains, and the somber visits from grieving families, but also by the emotional response of America as a whole, and the powerful new iconography that was associated with the disaster—these New York firemen as tragic heroes, these skeletal walls, these smoking ruins as America's hallowed ground. Whether correctly or not, the workers believed that an important piece of history was playing out, and they wanted to participate in it—often fervently, and past the point of fatigue. From the start that was the norm. There were some who could not stand the stress, and they had to leave. But among the thousands who stayed, almost all sought greater involvement rather than less.

The truth is that people relished the experience. It's obvious that they would never have wished this calamity on themselves or others, but inside the perimeter lines and beyond the public's view it served for many of them as an unexpected liberation—a national tragedy, to be sure, but one that was contained, unambiguous, and surprisingly energizing. Was this not war, after all? Probably it was, though at some early and willing stage of the fight: the workers believed wholeheartedly that they were righting a wrong, and that it was their duty to act quickly. The urgency of the job swept away ordinary responsibilities and the everyday dullness of family life, and it made nonsense of office paperwork and tedious professional routines. Traditional hierarchies broke down too. The problems that had to be solved were largely unprecedented. Action and invention were required on every level, often with no need or possibility of asking permission. As a result, within the vital new culture that grew up at the Trade Center site even the lowliest laborers and firemen were given power. Many of them rose to it, and some of them sank. Among those who gained the greatest influence were people without previous rank who discovered balance and ability within themselves, and who in turn were discovered by others. The unexpected ones were front-line firemen and construction workers, young engineers, and obscure city employees. Their success in the midst of chaos was an odd twist in the story of these monolithic buildings that in the final stretch of the twentieth century had stood so visibly for the totalitarian ideals of planning and control. But the buildings were not buildings anymore, and the place where they fell had become a blank slate for the United States. Among the ruins now, an unscripted experiment in American life had gotten under way.

The site roared twenty-four hours a day for nine full months and beyond. From autumn through winter and into spring the crews labored in twelve-hour shifts, got some sleep, and came back for more. The enormous scale of their workplace is difficult to convey. Under Mayor Rudolph Giuliani's insistence that New York return quickly to normal, the restricted zone had shrunk by late September to forty downtown blocks bounded by Chambers, Broadway, Rector, and the Hudson River—which was still a large area to be out of commission. At the heart of it, under the skeletal walls rising to 150 feet above the street, the debris spread across seventeen acres in smoldering mounds. It was dangerous ground, of course. Workers at the site called it simply "the pile." In detail the topography was complex, with strange craters and caves, unstable cliffs, and unexpected remnants of the World Trade Center as it had been before—the torn sculptural sphere on the ruined plaza, the amputated stores with their displays of goods no longer for sale, a row of bicycles still securely chained to a rack, a lamppost here or there still standing.

Passersby peering at the ruins from afar, from beyond the perimeter fences along Broadway or Chambers, sometimes expressed amazement that two 110-story buildings could collapse into a mass that was relatively so small. But there was no mystery to the dimensions. Because the Twin Towers had been as much as 90 percent air and 10 percent structure, they had contained the equivalent of approximately eleven solid stories of steel and concrete—and beneath these uneven mounds, which stood five stories high and spilled into the streets, their remnants filled the foundation hole to bedrock, six levels down.

Indeed, from the workers' close-in view, the ruins accounted for the former buildings all too precisely. It became apparent that the initial release of powdered concrete into the air, however large it had been, had amounted to an insignificant proportion of the total weight, and that almost every pound of the original structure remained to be extracted, inspected, and hauled away. So the ruins were not small at all. On the scale of people at the site they were in fact gargantuan—not mounds but mountains that reduced the expeditions venturing out across them to ant-sized columns traversing steel slopes. More than 1.5 million tons of heavy steel and debris lay densely compacted there, tied together like steel wool and complicated by the existence of human remains.

The weight alone defied imagination. What does a chaos of 1.5 million tons really mean? What does it even look like? The scene up close was so large that no one quite knew. In other countries clear answers would have been sought before action was taken. Learned committees would have been formed, and high authorities consulted. The ruins would have been pondered, and a tightly scripted response would have been imposed. Barring that, soldiers would have assumed control. But for whatever reasons, probably cultural, probably profound, little of the sort happened here, where the learned committees were excluded, and the soldiers were relegated to the unhappy role of guarding the perimeter, and civilians in heavy machines simply rolled in and took on the unknown.

The most effective of these machines were giant diesel excavators—tracked monsters with hydraulic arms and grappler claws that could tear tangles or close around sections of heavy beams and worry them clear of the pile's embrace. They were voracious, incautious, ignorant, and unstoppable. Soon there were dozens of them in action, painted orange or yellow, working in tandem to "daisy chain" the debris down to the trucking operation in the streets, straining under their loads, with smoke and dust rising around them, and flames licking their claws. For months this was how the unbuilding went—like a dance of dinosaurs in a volcanic land of steel, with men there too, moving through, restlessly searching for their kind. The effects were astonishing. Though the enormousness of the ruins made the final goal of "a clean hole" seem remote and difficult to imagine, the surface features of the pile changed so fast that time itself was compressed, and memories just a week old grew faint. People got caught up in that. Many of the core group were camped in hotels or temporary apartments nearby. The smart ones reminded themselves that they had recently led ordinary lives, and that they would return to them. But weeks felt like months, and months felt like years, and as winter came, it was ordinary life, and not the site, that began to seem dreamlike and far away.

Early on I found a piece of high ground from which to watch the changes taking place. It was inside the severely damaged and deserted Bankers Trust building, a black steel structure forty floors high, which stood across Liberty Street from the ruins and was eventually draped in dark safety netting and hung with a large American flag. In 1999 the German company Deutsche Bank had absorbed the Bankers Trust Corporation, and with it had acquired this building, whose offices it had occupied until the attack. During the South Tower's collapse steel spears and column sections had plunged into Bankers Trust, tearing a huge gash in its north face, destroying a load-bearing column for ten floors, spilling tons of office innards, and leaving the partially demolished floor slabs to sag like hammocks over a deadly void. In a crater at the base a mound of rubble lay laced with the remains of people who had been killed in the South Tower or on the street. There was serious concern at first that the building would not stand, but it did, and sturdily, because of redundancies in its design. The back offices, away from the Trade Center, were fine. Apparently only one person had died inside. Firemen checked the empty spaces quickly, leaving their fluorescent-orange graffiti—SEARCHED—on each floor. In the dust that coated one wood-paneled wall someone, I assumed from the Boston Fire Department, drew a sad face and scrawled,

> Kill All Muslims
> 9-11-01
> B.F.D.

Then for a long time the Bankers Trust building was left alone.

I went there first one afternoon, and climbed a broken escalator to a ruined entranceway that was lit through blasted walls and shattered windows, and strewn with rubble. The air inside was hazy with smoke from the fires across the street. Wearing gloves and a rubber respirator mask, I stirred through the rubble like an archaeologist on contaminated ground, searching for traces of its former inhabitants—in this case the bankers of just a few weeks before. The search was disappointing, because the forces of destruction had swept the entranceway clear of their presence. But then I climbed a dead dark stairwell, and several floors higher emerged into a scene richly preserved from their lives.

It was the Deutsche Bank executive dining area, torn open at one end and covered in the gray Trade Center dust, but otherwise intact—a narrow hallway, hung with art, giving onto a string of small private dining rooms that were acoustically damped and intended to facilitate privileged conversation. In a foyer, by a vase of plastic flowers, a menu dated September 11, 2001, showed the morning's breakfast: smoked-salmon omelets and chocolate-filled pancakes, among other offerings. The business cards of the host, a man named Mohamed Idrissi, lay on a countertop next to a telephone and a closed reservation book. I explored the hallway, moving carefully out of respect for the museumlike quality of the scenes, some of which, judging from the pristine condition of the dust, had remained untouched since the Trade Center's collapse. Each little dining room was furnished austerely with a table and four straight-backed chairs, a button for summoning service, and a telephone. It was clear that most of the rooms had been unoccupied at the time of the attack. They were set for lunch—with fresh place mats, plates and utensils, and sets of stemmed glasses, some of which had been capsized and broken by the pressure waves and lay now as they had fallen, like everything else here, under a feathery gauze of the Twin Towers' remains.

Down the hallway I discovered a dining room that had been in use, evidently by two people, whose breakfast still sat on the table as they had left it. At the center of the table was a platter of fruit—oranges, apples, and a stem of withering grapes. Beyond it lay the remains of the meal in progress—an omelet mostly eaten, a bowl of cereal mostly not, pastries in a napkin bed, two nearly finished coffees (one creamy, one black), two types of sugar cubes (both brown and white), a toasted English muffin, four little unopened jars of jam, a display of artfully overlaid butter patties, and two glasses of orange juice, half drunk and now darkening with age. The former occupants had a ghostly presence that was hinted at by the patterns they had left behind. They seemed to have maintained their poise at the time of their escape. The silverware was laid carefully aside, the napkins were on the table, the chairs were pushed just slightly back. It looked as if they had slipped politely away. It appeared to be a mutual thing, as if they had encouraged in each other a sense of decorum and calm.

The opposite was true at the end of the hall, where in a conference room I came upon a scene of mass panic and flight. It was the remnant of a breakfast meeting for about twenty people—a U-shaped array of tables facing a display board, and a varied buffet with the food looking almost still edible. Somehow the attendees must have known that the evacuation was no drill. In their haste to escape they had abandoned their briefcases and overnight bags, and even dropped some car keys and purses. I poked through the detritus, knowing from the disturbed dust that I was not the first visitor to have done so. Indeed, it became obvious from the traces where laptops had been, and from all the opened bags, that the room had been systematically rifled for valuables—whether by errant firemen, policemen, or construction workers hardly mattered. All three groups were at various times implicated in a widespread pattern of looting that started even before the towers fell, and was to peak around Christmas with the brazen theft of office computers from this very building. The psychology involved was complex, and had more to do with the breakdown of the social order—a sense of crisis and of special privilege—than with any particularly criminal inclination to steal. People urinated on the carpets in Deutsche Bank as well. And presumably the same people who had robbed this room had also written "Fuck Bin Laden" on the display board, because a righteous sense of war was part of the package too.

But to me none of that seemed important anyway. Whether they had been violated or remained pure, the Deutsche Bank dining rooms were artifacts from an event that, within the inner world of the World Trade Center, seemed already to exist in distant history. Through the winter I kept returning to them, these set pieces from the past, and with every visit I felt satisfaction that they remained untouched. The pastries petrified. The juices slowly evaporated and, like the coffee, turned into black crusts. The fresh fruits fermented. For a while they were infested with gnats and other bugs, and then not. I never found out why the rooms were neglected, in part because Deutsche Bank responded to the September 11 attacks in a cowed Neither-we-nor-our-employees-will-comment style. The City of New York eventually shored up Bankers Trust for the safety of the streets below, but otherwise the building was left alone. No one was sure whether ultimately it would be torn down or restored; stories flew back and forth, and settled finally on a rumor about an insurance standoff over mold.

Fine. Bankers Trust was at its best as a ghost building anyway. You could use it to go back in time to a little bridge that spanned a sudden change in the story of the United States—not the grand "loss of innocence" proclaimed that fall in the press but, more modestly, a shift from an era of complacency preceding the attack to a period of creative turbulence just afterward. You could also then go upstairs, stand in the building's open wound, and look out at the turbulence in action across the magnificent smoking terrain of the Trade Center site—this ever changing American geography. If there were others standing beside you, the conversation would often be about memory. Do you remember when the pile was high, and then higher than now? When all the skeletal walls stood, or when one or another came down? When the upper road was built, or the lower road, or the first Tully ramp, or the second? When the Liberty Street wall sagged, or when the outline of the foundation hole became visible from above, or when we dove underground along the east wall by the escalator box—where is that now? Do you remember the day when there were twenty bodies, and the week when there were none? Or even, Do you remember the Twin Towers themselves? Can you really quite imagine them anymore?

The underground, beneath the pile, was a wilderness of ruins, a short walk from the city but as far removed from life there as any place could be. It burned until January, and because it contained voids and weakened structures, it collapsed progressively until the spring. From the start its condition dictated the nature and location of the work on the surface. Specific knowledge was necessary not only to protect the people on the pile, particularly from cave-ins, but also to maintain the integrity of the now precariously supported "slurry wall," a seventy-foot-deep concrete shell that enclosed the foundation hole to keep the tidal waters of New York Harbor from flooding in. The job of mapping the chaos fell to a small team of about six engineers who did some of the riskiest work at the site, climbing through the crevices of a strange and unstable netherworld, calmly charting its conditions, and returning without complaint after major collapses had occurred.

The engineers finished the initial surveys in a couple of months, which was fast, given the size of their domain—six levels of ten acres each within the foundation hole, and an adjacent two levels of six acres each in the Trade Center complex to the east, all lying in confusion. By mid-November only one important underground area remained to be explored—a place people called "the final frontier," located deep under the center of the ruins, at the foot of the former North Tower. It was the main chiller plant, one of the world's largest air-conditioning facilities—a two-acre chamber three stories high that contained seven interconnected refrigeration units, each the size of a locomotive and capable of holding up to 24,000 pounds of dangerous Freon gas.

Freon is a manufactured product containing chlorofluoro-carbons (CFCs), and its use has been restricted by international accord because of the damage it does to the atmosphere's protective ozone layer. The threat it posed at the Trade Center was more immediate, and stemmed from the fact that it is a heavy gas and it aggressively displaces the oxygen in the air that people breathe. With the huge quantities potentially involved here (as much as 168,000 pounds, under pressure, if the tanks remained intact), a sudden leak would fill the voids underground and spread across the surface of the pile, suffocating perhaps hundreds of workers caught out on the rough terrain and unable to move fast. To make matters worse, if the Freon cloud came in contact with open flames, of which there were plenty here, it would turn into airborne forms of hydrochloric and hydrofluoric acids and also phosgene gas, related to the mustard gas used during World War I. Then it would go drifting. Evacuation sirens were installed around the edges of the pile, but in tests they often failed. Evacuation maps were printed with elaborate routes and mustering points, but of course they were not read. People accepted the danger. The standard advice, "Just run like hell," was delivered with a little shrug. Everyone knew that if the Freon came hunting for you at the center of the pile, you would succumb.

So for two months the Freon lurked like a beast inside the ruins, hiding in its den deep among channels that were venting noxious smoke and steam. By mid-November it seemed that the diesel excavators, which were digging in a valley of ruins between the mounds of the Twin Towers, would soon be encroaching on the chiller plant from above. If in the natural confusion of the work one of them tore into a loaded refrigeration unit, or merely damaged a pipe or a valve, the effect could be catastrophic. In private conversation the engineers speculated that the chiller plant had been severely damaged by the North Tower's collapse, and that most if not all of the Freon had vented immediately into the rising column of smoke and dust. But they did not know this for certain, and could not gamble with people's lives. Moreover, now there was disturbing news that traces of Freon had been found in the water that was pooling just above bedrock seventy feet down, at the lowest (B-6) basement level of the foundation hole.

Since the water came primarily from the efforts to suppress fire and dust on the surface—from streams being played into the eruptions on the pile—it was possible that the Freon traces had simply been washed from tainted debris. Alternatively, the traces perhaps indicated a leak from one of the auxiliary tanks located in a less heavily damaged structure at the north end of the foundation hole—tanks that would soon be emptied by a specialized crew. But it was possible, too, that the engineers' speculation had been wrong—that the main chiller plant remained largely intact, and that the Freon inside it was somehow starting to escape. Either way, the time had come to take a look around.

In preparation for the effort, additional pumps were hooked up on the B-6 level, and over several days the flood in the foundation hole was drawn as low as it would easily go—to a depth of about one foot. The engineers knew that the route through the ruins would be tight and uncertain. A "spelunking run," they called it, and because they would go deep, a "B-6 crawl."

I was invited along, as I'd been many times before. The pile that day was burning heavily under a south wind that spread ash and the sweet smell of the ruins far into the city, eliciting complaints. The sky was gray with autumn clouds; the air was soft and cool. Amid the roar of machinery we met at a plywood shack near the southwest corner of the ruins—twenty men in helmets, hardhats, and high rubber boots. The shack was a doorway to the underworld. It covered the top of an old stairwell that descended into the ground within the so-called South Projection, one of two protrusions under West Street, where the foundation hole had been extended toward the Hudson River to encompass the PATH tubes. Since September 11 the South Projection stairwell, like that of the North Projection, had provided important access to the ruins below. For the first few weeks it had been choked with rubble, and had required careful negotiation, like a steep and unstable switchback trail. By now, however, it had been cleared, and could be taken without effort, twelve flights down to the abandoned tracks.

At the bottom, in the dungeon light from an overhead crater nearby, we collected in the mouth of the PATH tube for a brief discussion. The air smelled dank and musty, in a fading combination of train brakes, oil, and steel. Some of the people were new to the scene: a police rescue team of five men, looking dandyish in their kneepads, carabiners, and helmet lamps, and an equal number of firemen, less proudly dressed, who had recently arrived for the standard one-month rotation. The firemen were young, and visibly more relaxed than the police. Several ventured like sightseers into the PATH tube, playing the beams of their flashlights across its iron rings, into the green and red puddles of oily fluid, and down the rusting tracks that stretched westward into the blackness under the Hudson River. In about two hours you could walk through the tube to New Jersey and back, as people sometimes did, if only for the solitude. But today the job was serious, with risk involved, and it lay in the other direction, northeast across the complications of the ruined basement and into the unexplored.

The expedition carried several emergency air packs and an electronic sniffer to warn of low oxygen and the presence of Freon—but none of this would help in a collapse, which the engineers still believed was the greatest danger we faced. It was reassuring, therefore, that the group included underground veterans—indeed, some of the Trade Center's most experienced men. These were people unheralded on the outside, several of whom in retrospect now stand among the greats in the recovery effort. One was a watchful and laconic fireman named John O'Connell, a specialist in building collapses, who had served as an example to the crews from the very start, calmly smoking cigars in even the most threatening conditions, and providing people with a necessary model of skepticism in the face of all the hype.

Beside him stood Sam Melisi, another fireman, whose talents turned out to be even more valuable. He had a gentle and self-effacing manner, which seemed out of place at first but stemmed from profoundly altruistic impulses that over time became widely recognized, eventually giving him a moral authority that no one else at the site came close to matching. That authority translated into the power to make suggestions that others were willing to follow. The power surprised and plagued him to the end; he did not think of himself as a leader, and in other circumstances he probably would not have been one. Nominally he was always just a fireman, a collapsed-building specialist like O'Connell, with front-line duties that included assisting the engineers on these underground runs. But already by mid-November some people had started calling him "Saint Sam," as in "Saint Sam the Fireman." He rose to it uncertainly, but gradually assumed the all-important role of mediator.

The French flew him to Paris for a day to plant a tree in the Luxembourg Gardens, as if he were just another New York "hero," good for a photograph and a night out on the town; they had the luck of the draw and never knew it. Melisi remained unknown to New York, too, which had nearly as distant a view of the site, and had actually been surprised by the outbreak of fighting there between firemen and the police on November 2—a power struggle that came to the surface after weeks of mounting tensions. Afterward the political operatives at City Hall must secretly have believed that they maintained the peace between the multiple opposing forces on the pile, but in truth, if anyone did, it was Sam Melisi.

Of all the people setting out now for the chiller plant, twenty men redefined by these ruins, the one who would have the greatest influence on the unfolding story was an obscure engineer, a lifelong New Yorker named Peter Rinaldi. At age fifty-two, Rinaldi was an inconspicuous olive-skinned man with graying hair and a moustache, who observed the world through oversized glasses and had a quirky way of suddenly raising his eyebrows, not in surprise but as a prompt or in suggestion. He had grown up in the Bronx as the son of a New York cop, had gone to college there, and had married a girl he had met in high school. Though he and his wife had moved to the suburbs of Westchester County to raise their three sons, he had never cut his connection to the city, or ever quite shed his native accent. For twenty-eight years he had commuted to the World Trade Center, to offices in the North Tower, where he worked for the Port Authority of New York and New Jersey, deep within its paternal embrace and completely secure in his existence. There was an early warning in the terrorist bombing of 1993, an attack that trapped him in an elevator. Nonetheless, he was wholly unprepared for the destruction that followed in 2001.

The institution for which he worked, known simply as the Port Authority, is as much an empire as an organization—an insular and enormously wealthy bi-state agency, exempted from ordinary governmental constraints, that was created in 1921 to build and manage the area's transportation facilities, and was given the right to finance itself by issuing bonds and collecting its own tolls and taxes. It operates every major airport in the area (Newark, La Guardia, Kennedy, and Teterboro), manages all the ports (including one of the world's largest container facilities, at Newark), runs a busy railroad (PATH), owns New York's commuter-bus stations, controls many of the city's lucrative tunnels and bridges (among them the Lincoln and Holland Tunnels and the George Washington Bridge), and even has its own private army—an independent police force 1,300 strong. More to the point, it also built, owned, and until recently operated the World Trade Center, where it maintained its headquarters; on September 11 it occupied space on thirty-eight floors and on each of the six subterranean levels. All of that was lost when the buildings collapsed, along with the lives of seventy-five Port Authority employees, some of them maintenance men who could have escaped after the South Tower fell, but went grimly back underground to find their co-workers. It would be hard to overestimate the ensuing sense of exposure and uncertainty for their colleagues who survived. During the days after the attack, when to New York City officials the Port Authority seemed to have disappeared, it was hunkered down across the river in its New Jersey offices, suffering through a collective emptiness so severe that people themselves felt hollowed out.

Peter Rinaldi felt it too, though he was far away at the time of the attack, vacationing with his wife, Audrey, on the Outer Banks of North Carolina. The North Tower was like a home to him. He'd had a corner office on the seventy-second floor, and a satisfying job there overseeing the engineering for all the Port Authority bridges and tunnels. He especially liked the George Washington Bridge, and had once appeared on the Discovery Channel to praise its resilience. Recently he had found a million dollars in spare funds, and without going through all the usual bureaucratic rituals had spent it on decorating the span with permanent lights. Others in the Port Authority had criticized him afterward for bypassing procedures—but only in private, and as friends. Rinaldi took it in stride, and never felt threatened, because the atmosphere at work was always so stable.

But on the Outer Banks, on television, he watched everything come apart. It took merely two ten-second pulses. Months later he was still struggling to describe the shock. To me he said, "It was surreal. You're on vacation, and you're looking at this, and it's a place where . . . I knew all the people there. It's a place you're supposed to go back to work, your office. It's everything you do. You just watch this thing collapse. It was like, you couldn't really believe it, in terms of all this loss. You say, 'That's not really real. This can't be gone.' And then you think, Well, what's going to happen afterward? What are you going to do? How are you going to do this? Where are you going to be? What's going to happen with the agency? How many people were lost? It could have wiped out our whole two thousand people we had in that build-

ing. Or the whole organizational structure. The whole center of
the PA was in that building."

The whole center of his life. He said little at the time, and
spent another day at the beach, but the event had devastated
him. His wife told me that he became a man she had never seen
before—preoccupied, passive, withdrawn. Part of it was that he
knew so many people who had died, several of whom were close
friends; back in New York he went to their wakes and funerals.
But another part of it was that he had been stripped of his shelter, irrevocably, and that, as he told me later, he doubted his ability to proceed in a disorderly world.

And then, adding to his uncertainty, having briefly reunited
with his old staff at temporary quarters in New Jersey, Rinaldi
was reassigned to New York City's recovery team—a small group
of strangers to him who were camped in the kindergarten rooms
of Public School 89, near the ruins, and, in the midst of the most
disorderly conditions imaginable, were improvising a response.
These were primarily Kenneth Holden's people from the DDC.
Rinaldi was given the job of supervising the consultants who had
been brought in for the specialized belowground engineering. I
met him at the start, when he seemed tentative and out of place,
and I watched him through three seasons to the end, by which
time he had become, both above and below ground, the one man
everyone turned to for an opinion.

The attempt on the chiller plant happened about a third of the
way through his odyssey. Rinaldi was nominally in charge, but
he neither gave commands nor asserted himself by taking the
lead. I think by then he had already discerned what some others
never did—that the imposition of conventional order on these

ruins was a formalism or a fiction, and unnecessary. Progress was made instead in the privacy of a thousand moments, on loose, broad fronts, by individuals looking after themselves and generally operating alone. This was how we turned now and drifted toward the jumping-off point for the unknown, twenty men, each picking his path.

Under the crater and by the light of the sky far above, Rinaldi and I paused by a pile of rubble that we had occasionally crossed before, and we discussed the news that a fireman's body had recently been found there, buried under a thin layer of dust and debris. It seemed strange that for two months neither we nor anyone else had noticed him. Rinaldi mentioned that in the chiller plant—should we find a way to it—we might encounter the body of a certain Port Authority employee, a man who was believed to have sheltered there. So be it. After so many other encounters at the site, the prospect elicited neither hope nor horror. People grow used to the dead.

Beyond the crater our route was lit by a string of utility lamps installed for the pump crews that had been drawing down the basement flood. We gathered where the pumps were running, on a concrete platform above an area of black water. John O'Connell, in a wide-brimmed fire helmet, slouched by a steel column, smoking a cigar and looking typically unimpressed. The police seemed to think that their sergeant was in charge, and apparently the sergeant did too. He announced that he would be probing the water with a steel pipe, and he advised everyone to stay in line close behind him. No one quite looked at him, or responded. The sergeant warned about ten-foot sumps now hidden within the flood, into which a person could disappear and drown. Sam Melisi checked a map for the specific locations, which indeed were shown near the chiller plant. Rinaldi reminded us of our schedule. One by one we stepped off into the flood, and with flashlight beams probing the huge forms of ruined structures, we waded in a single long file into the darkness.

The sergeant at the lead soon gave up trying to sweep water with his pipe and began to use it as a crutch to steady himself as we moved across submerged debris. Each man felt his own way, not walking so much as sliding forward experimentally, sloshing through the opaque water and taking care not to trip over fixed objects or to step heavily on the looseness underfoot. It was clear that anyone falling here risked injury and, if cut, a serious infection. Nonetheless, our progress was fast, and it soon reduced the view of the pumping platform behind to a distant light, and then to nothing at all. We moved across a cavern bordered by massive collapses and traversed by man-sized pipes, one of which, labeled "River Water" and known to be closed off, was the main intake line for the chiller plant. Whether because of nearby fire or ongoing collapses, the air turned thick with dust. With the exception of the ever dismissive John O'Connell, we donned our respirators, limiting conversation. The cop with the electronic sniffer was clearly concerned, and was checking the indicators so often that he began falling behind. He lifted his mask and called, "Oh-two's okay! All gasses okay!" We slowed to let him catch up. The darkness grew loud with the sound of falling water, which turned out to be bands of a mysterious subterranean rain—too heavy to result from the firefighting on the surface—falling from the confusion of ruin overhead. We moved through the rain, and fifty yards farther on climbed onto the dry ground of a catwalk built above an area of heavy machinery. We paused where the catwalk ended, in latticework stairs leading down into the flood-water on the far side. By then we had progressed deep into the pile, and by ordinary measures were only a short distance from the main chiller plant. There was still no indication of Freon. The ruins, however, were closing in oppressively, with crazily angled slabs tilting down into the waters ahead amid a confusion of rubble slopes that obscured the blackness beyond, where an access route might possibly be found. The place looked like a trap, and dangerous as hell. The police apparently thought so too; they bunched on the stairway and tested their radio link.

Rinaldi discussed the situation with a couple of Trade Center facilities men who had come along because of their expert memories of the buildings. Given the extent of the destruction that clearly lay ahead, it seemed unlikely that any of the chiller plant could have survived. In that sense, the work here might already have been accomplished—and there was certainly reason to be skeptical about the benefits of moving forward. At the same time, if for no other reason than the draw of the unknown, Rinaldi was obviously unwilling to retreat. Within the private culture of the Trade Center underground this was understood without explanation: it was simply very difficult to turn away from space that remained unexplored. Briefly, therefore, we were at an impasse. But then one of the engineers took matters into his own hands, and without saying a word descended the last steps into the water and set out alone. The police had given up any pretense of command. No one called the engineer back. Rinaldi watched him for a while—a tall, thin silhouette familiar to us all, pushing steadily into the ruins—and then he and I and most of the group set out too.

The tall, thin silhouette belonged to a man already famous for his independence. He was Richard Garlock, age thirty-three, a boyish-looking structural specialist who worked for the Trade Center's original designer (Leslie E. Robertson, the "engineer of record") in a small company called LERA, which for thirty years had served as the main Trade Center consultant to the Port Authority, and which had its offices nearby. LERA was known at the site less for its history than for its current feeling of having been marginalized during the cleanup efforts under way. Certainly it had responded valiantly to the attacks of September 11: Garlock and others from the firm were running toward the Trade Center to provide a structural assessment of the North Tower's wounds when the second airplane came in and hit the South Tower. They had returned to their offices and were volunteering their expertise to the Fire Department when the towers collapsed. Their firm had taken the lead after the terrorist bombing in 1993, and it was only natural to assume it would do the same now. Instead, when they made the offer to help, they found that LERA had been relegated to a minor role—to be called on by other consultants as needed. Understandably, they took this as a slight. They never said so publicly, but on the level of their front-line engineers it was clear. I repeated their tale of woe one evening to Audrey Rinaldi, who had arrived for dinner downtown from her job as a nurse and hospital administrator in the Bronx. She said jokingly, "You know, engineers are human too," as if the point could be debated. But there was always plenty of proof at the Trade Center site.

The prime consultant on the job was a large and well connected New York firm called Thornton-Tomasetti—the engineers of the Petronas Towers in Kuala Lumpur, and of the United Airlines terminal in Chicago, among other monumental structures. With its 400 employees and enormous institutional expertise, Thornton-Tomasetti could have ignored LERA's feelings entirely but for one important catch: since most of the complex's technical drawings had been stored (and thus destroyed) within the North Tower, LERA had the only complete and readily available set of the original plans—and it was careful to hand over just what was needed. In fairness, LERA's position probably had more to do with self-protection than with pique. The firm was badly exposed to the waves of lawsuits already coming into view, and it needed to maintain some sort of control. For similar reasons it wanted to make its own detailed assessments of the ruins. People accepted this, however grudgingly. LERA had been smart if not supple. Without withholding information or being obstructionist, it had used the technical drawings adroitly and had found a place at the center of the operation. On all the important surveys now, including this chiller-plant run, Garlock was the man holding the blueprints. And having fought so hard to be here, he wasn't about to retreat or to leave these ruins unexplored.

Actually, Garlock was fearless anyway. He was a moralist at times, but an affable fellow too, with a distracted graduate-student manner that contrasted pleasantly with the harshness of the surroundings. I found it interesting to watch him work—all the more so after I realized that he was not engaged in some sort of geekish theater (my first assumption) but was genuinely absorbed by the minutiae of even the most obscure corners of the ruins. When I described his enthusiasm to Audrey Rinaldi, who had taken it upon herself to explain engineers' personalities to me, she acted unsurprised.

I said, "But honestly, when you're deep underground and everything is fine, just how fascinating can it be that this is Panel D-15 or A-5?"

She looked at me pityingly, as if she hated to shatter my illusions. She said, "I keep telling you . . ."

But Garlock seemed unusual even to the other engineers. Some of them started calling him "Planet Garlock," as if he had descended from a faraway place. He would go wandering off through the subterranean ruins, gazing intently through small spectacles at the columns and beams, often with the hint of a quizzical smile, making notations on his cherished blueprints, with apparently only a vague awareness of the danger signs around him—the jolt of a collapse far below, the rattle of cascading debris, the ominous groaning of weakened structures overhead, or, in the early days, the streams of molten metal that leaked from the hot cores and flowed down broken walls inside the foundation hole. The others would exchange looks and ask Garlock to hurry. In my experience he never did. He was interested in every part of the ruins, but really relished the underground.

The truth is, we all liked it there, if not to the degree that Garlock did. The underground was the Trade Center's inner realm, the destination at the end of a progression that with each step took people closer to the heart of the ruins: across the perimeter, to the edge of the pile, onto its surface, and then down into its insides. It was also simply an intriguing place to be, and less macabre than outsiders supposed. The dead lay nearby, but for the most part they were buried in the inaccessible cores—debris so tightly compressed that even the rats, said to be arriving from all over downtown, were unable to burrow in (or so we told ourselves). The cores were shaped like inverted mountains, or cones, that grew narrower with depth. They filled the towers' footprints to bedrock and occupied the center and south end of the foundation hole. They were bounded by looser ruins and basement structures in an arc from west through north to east. And it was among those looser ruins, after Garlock spurred us into motion, that we continued now to explore.

Garlock roamed south into the darkness among the piles of debris. Others roamed north. Rinaldi and I moved slowly to the east. In an area of mountainous collapses we came eventually to another catwalk, which we climbed until it ended in a slope of loose rubble. There was nothing but blackness behind and ruin ahead, and no sign or sound of the others. Blocks of concrete the size of cars hung over our heads, one dangling on rebar.

Through the respirator mask I asked Rinaldi, "How stable do you think that is?"

He said, "It's not."

I asked, "Have you noticed how guys will pretend not to notice?"

He said, "Yeah," and we both laughed. But we knew that if it hadn't fallen yet, it wouldn't fall now.

Almost by chance then Rinaldi's light picked out a man-sized hole in the debris slope, above us and to one side, and through it, faintly illuminated, the form of a rounded mass of steel. It was heavy equipment of some kind—and given its location, it had to be inside the main chiller plant. Rinaldi crawled up and in, spilling dust and broken concrete down the slope, and I followed close behind. The space in which we found ourselves gave the impression of an oversized version of a ship's engine room, but reduced and hemmed in by massive collapses, and lying in complete ruin. There was no sign of the missing Port Authority man and, though we lifted our masks to check, no smell of his death. We squeezed between a partially crushed tank and its battered pumps, and were barred from further progress by a wall of broken slabs and twisted steel. After a while one of the Trade Center facilities men arrived, followed later by Garlock and the young fire crew. For no good reason one of the firemen tried to climb a twenty-foot chimney in the unstable rubble, and he had to be pulled back and told to calm down. No one thought much about it, because so many of the firemen were emotional, especially when they were new. Rinaldi took pictures of the ruins. The damage was even worse than the engineers had expected it to be, and it offered the first hard evidence of a fact that was eventually verified during the excavation from the surface—that the Freon had vented, and the beast at the center of the pile was a myth.

At its most chaotic the underground was like the abstract nether-world we encountered during the chiller-plant run—as shred-ded as the surface of the pile, yet without the organizing principle of the sky. But that was the extreme. Much of the un-derground was intuitively easy to understand. It consisted of parking garages, often in some stage of collapse, where more than a thousand cars now stood abandoned and covered with the standard gray concrete dust. A disproportionate number of the cars were BMWs, Jaguars, Lexuses, and the like—indicating, if nothing else, the preponderance of a certain culture that had thrived here. Although a few seemed strangely untouched, most were crushed, sliced, blasted, or burned. Along the north side, where the basement structure remained strong and intact (and was ultimately preserved), the fire had been so intense in places that it had consumed the tires and interiors, and had left hulks sitting on axles above hardened pools of aluminum wheels. Three presidential limousines stood in there too, but they were locked away, and remained unscathed. When access was opened, the Se-cret Service rushed in and with a great show of secrecy loaded the limousines onto flatbed trucks, covered them with tarps, and hauled them back to Washington.

There was a romantic idea, widespread at first, that the Trade Center underground would contain wonderful and varied trea-sures—but it never quite panned out. Along the north side be-fore the collapses there were firing ranges and gun rooms for various police agencies, and vaults containing confiscated narcotics and cigarettes, and there was a collection of artifacts from a Colonial-era African-American burial ground, but aside from the garages and the PATH facilities, the basements consisted for the most part of utility spaces and storage rooms in which the things that were kept were basement things—tools, wire spools, spare chairs and partitions, and, in one particularly claustrophobic corner, glossy brochures from the former 107th-floor observation deck of the South Tower. Out beyond the foundation hole, in the burned-through remains of Building Five, a Citibank vault was stacked with a fortune in bundled currency—which, however, had been baked and turned to ash. Nearby, a vault left open at Morgan Stanley turned out to be almost as disappointing. It contained $2.7 billion, but in financial certificates that would have been nearly impossible for a thief to translate into cash. A Brinks car reportedly holding $14 million was stuck in the ruins for a few weeks, adding an exotic touch to the underground until Brinks was able to remove it. The newspapers reported the loss of government safes containing top-secret documents, for what that was worth. They also reported the loss of important legal documents, news of which caused no stir.

Ultimately, only one Trade Center treasure was worthy of the name. It was a hoard of gold and silver ingots, valued at about $250 million, that filled a two-story vault in the remnants of an old railroad station under the ruins of Building Four.

A quarter of a billion dollars' worth of gold and silver turned out to be a lot of ingots—more than 30,000 in this case. The ingots weighed up to seventy pounds each, and were stacked on wooden pallets that could be moved within the vault by forklift and internal elevator. Their total weight was 1.9 million pounds. The vault belonged to the Bank of Nova Scotia. The bank held the treasure to legitimize trading in the precious-metals market. In practice it was expected that the metal would remain in the vault, even as it was bought and sold in various forms. There were guards to make sure that nothing went wrong with the stock. It never did. Even after the attacks of September 11 security was not a worry, because the guards had bolted the door before fleeing, and mounds of heavy debris blocked any conceivable access. The treasure lay locked in a vault inside a vault. At least that was the thinking.

But when, at the end of October, a pathway was finally cleared, down a truck ramp from the north and through an old railroad tunnel, the bank's initial entry team discovered that others had been there before, attempting to pry open the vault's door and to cut in from above, in both cases unsuccessfully. Though it was presumed that the intruders had been construction workers, it never became clear who exactly they were, where they had come from, or how they had proposed to get away through these ruins with more than just a few ingots. But if the unbuilding of the World Trade Center had already shown one thing, it was that the workers there were resourceful and persistent. The bank hurriedly organized a convoy of armored trucks and over the course of several days, amid another display of pomp and secrecy, moved the gold and silver out. It took 120 trips. No one was supposed to know that the destination was the Brooklyn Navy Yard, but everyone did. Over the weekend people made a point of getting past the security detail—cops in bulletproof vests, cradling riot guns—and down into the vault to watch the treasure disappear. It was an innocent distraction. The vault inside was not merely undamaged but well lit and pristine. For one weekend it served the underground as a tourist destination.

But the most popular place down there was always the two-story PATH station, which stretched along the eastern wall at the deepest levels of the foundation hole, and was the most public of the Trade Center spaces that remained at least partially intact. Although it had suffered heavily, and continued to collapse throughout the fall and winter, it was considered to be relatively safe. In the last days of September, Peter Rinaldi pioneered the early route in, down the North Projection to the PATH tube and then by rubber raft along the heavily flooded tracks, a River Styx flowing deep under the ruins of Building Six. It was Rinaldi's first underground exploration. Richard Garlock went along, as did two other young engineers, who worked for the famous foundation firm Mueser Rutledge and played important roles at the site—a huge and garrulous man of Ecuadorian origin named Pablo Lopez and his wry, cerebral partner, an ex-carpenter and Columbia graduate named Andrew Pontecorvo, who was small by comparison and did some tough duty crawling into tight spaces. They were accompanied by a typical phalanx of firemen and cops.

They had three rafts, two of which were leaking air and on a couple of occasions had to be reinflated. The men pushed and paddled the rafts for forty minutes up the railway, eventually through widening spaces as the tracks branched into the PATH station. The water came up almost to the level of the station platforms, which remained dry. The men disembarked, grounded the rafts, and began to explore. Rinaldi was the one most familiar with the place, but he moved through it as if in a dream, noticing the complete silence and the little clouds of dust that rose in slow motion with every step he took. The darkness was so deep that it seemed to defeat his flashlight beam. A ghostly train stood there in an image of abandonment—the first cars crushed under heavy slabs, the remaining ones intact but empty, with their doors ajar.

Rinaldi knew already that no one had been inside on September 11—the train had been sitting idle in the station all morning, and it had simply remained. Two other trains, full of passengers, had left New Jersey and were bound for New York when word of the first attack came. In one of the more elegant moves of the morning they did not stop but with typical Port Authority aplomb continued to roll smoothly through the Trade Center underground and, without rushing, slipped back beneath the Hudson to safety.

Rinaldi had the wit to appreciate that sort of thing, but he wasn't in the mood for it now. Others on the team had started into the technical work of the survey—Garlock with his close-up inspections, Lopez and Pontecorvo with more distant views—that ultimately, after many more surveys, would result in a decision to restore PATH service quickly and reuse this station while a new permanent one was built. But that, of course, was the future, and for now Rinaldi was thinking more about the past. He climbed a dormant escalator to the station's upper level, and for old times' sake continued to take in the sights: the melted plastic signs, the pay phones on pedestals, the turnstiles, the candy kiosk still offering sweets, and then the part of the station that became a photographed favorite—the prominent Commuter Bar, with its bottles of booze, its racks of inverted glasses overhead, its open Heinekens on the counter.

I could never really appreciate that place, which to me suggested the style in which too many people here must have lived—stacked in their fluorescent-lit towers, removed from the city outside, and wanting to drink in the morning. Rinaldi himself was not much of a drinker, though he liked a good Chianti with dinner. But during another survey of the PATH station, when I mentioned my dislike of the Commuter Bar scene, he grew upset and said he disagreed, maybe because he still thought of all these ruins in some way as his home.

Rinaldi was human but also always an engineer: even during the worst moments of September 11 he continued to think analytically about the buildings. He was two days into the vacation on the Outer Banks, having breakfast with Audrey and friends, when his grown son called from Connecticut and said, "Hey, Dad. A plane just hit the World Trade Center." It had happened at 8:46, just a few minutes earlier. Rinaldi turned on the television and saw the first pictures: it was the North Tower that had taken the hit, high on its north face and apparently well above the engineering quarters on the seventy-second floor, where Rinaldi's staff was already at work. From the quantity of black smoke pouring out, Rinaldi realized that a major fire had erupted. Television reported that the airplane was believed to be small, possibly a commuter airliner. But Rinaldi knew the building's true dimensions, and he was not fooled by television's miniaturization. He said, "Boy, that's pretty large damage for a small plane." The entry wound seemed to stretch nearly the width of the North Tower's 209-foot face, and against that scale it formed the head-on silhouette of a very large airplane, banked to the left. Later, when it became known that the airplane was a Boeing 767, it was easy to discern even the outline of its low-slung twin jet engines. Some people were reminded of the perfect holes left in walls and clouds in cartoons. Rinaldi was more practical. He interpreted the tightness of the gash's pattern as a sign of the structure's admirable redundancies: the impact had severed much of the North Tower's all-important exoskeleton, the closely spaced steel columns that shared with an internal elevator core the work of supporting the building's weight, and yet there was no sign of further unraveling. Rinaldi thought about the weather, which was bright and clear, and he wondered how any pilot on such a morning could have made such a mistake.

But then, at 9:03, the second 767 came in fast from the south and, hooking a hard left turn, slammed into the South Tower. It hit lower than the first one had, but with much the same effect. Rinaldi saw the impact, the fireball, and the ensuing flames. One of the friends cried out, "My God, we're under attack!" It was true enough, but Rinaldi was concentrating on the immediate practicalities. He said, "They'll be evacuating down through the stairwells, and then they'll have to put these fires out." And that was true too. Rinaldi knew about the improvements that had been made in the stairwells since the terrorist bombing of February 1993, primarily in ventilation and emergency lighting, and he trusted that his staff in the North Tower would be able to reach the street from the seventy-second floor in about fifty minutes. If they had left the offices immediately, they would be almost halfway down by now, mixed into the crowds going steadily around and around, and probably starting to encounter firemen moving up to fight the flames. Even if they were descending more slowly, or for some reason had lingered in the office, there was probably no reason to worry about their safety. Rinaldi himself had needed five hours to escape the building when he'd been stranded there in 1993—and though today's attack was clearly more severe, this time the fire was located higher than the engineering offices, and the smoke was rising away. Rinaldi was of course horrified by the scenes unfolding above the impact zones, where people had broken out the windows, and some were clinging to the outside and jumping and falling away. But it can be understood if he reacted to these first minutes of the disaster like an old Port Authority hand, with more confidence in procedure than in initiative and speed.

He was not the only one. The memory of the 1993 bombing turned out to be a trap for many people on September 11, because it boiled down to the fact that though six people had died, the buildings had survived. The Fire Department was particularly proud of its reaction, which it promoted within the ranks as an important success story. Until September 11 it was the largest operation in the department's history—a sixteen-alarm response that initially occupied nearly half the force, and engaged specialized crews (which included John O'Connell and Sam Melisi) for twenty-eight days afterward. As is well known, the bomb was set by Islamic militants, who were easily caught and convicted. It was a powerful device, the equivalent of more than a thousand pounds of TNT, detonated by a timer just after midday in a yellow Ford Econoline van parked on the B-2 level of the underground garage, near the foot of the North Tower. It blew a crater through all six levels of the basement structure, rupturing the surface of the Trade Center's open-air plaza, destroying a hotel meeting room, and disabling the complex's "communications center," while also blasting downward through the stacked garages, lighting cars on fire, and dropping slabs into the chiller plant, causing peripheral damage. One of the people killed was a maintenance man who had placed his desk in an alcove in the North Tower's exterior foundation structure. He was handling paperwork there when the floor dropped out beneath his chair. For weeks afterward the desk remained exactly as he had left it, cluttered with his work but perched impossibly on a steel spandrel high above the crater, as if some crane operator with a dark sense of humor had put it there.

Rinaldi was late for lunch that day, and he was riding an elevator from his offices to the Port Authority cafeteria, on the forty-third floor, when mysteriously the elevator stopped. There were nine passengers aboard, including several other Port Authority engineers who knew the building well. Because elevator failures were not uncommon at the Trade Center, no one seemed frightened at first. The standard procedure was to push the red emergency call button; in theory, someone in the communications center would answer, and then find a way to work the elevator down, or send a rescue crew. This time, however, all that came back was a taped response, a voice repeating, *Your call has been registered—we will get back to you in a minute. Your call has been registered—we will get back to you in a minute.* During the nightmarish events that followed, the message played on for at least an hour, defeating one man's attempt to silence it with a bundled coat and adding a surreal quality to the experience— something like being trapped and tortured inside a voice-mail machine.

For a while the elevator was well lit. The engineers analyzed the problem, as engineers like to do. When a thin, acrid smoke began to filter in, they discussed the possibility that the elevator motor had shorted and burned; when the smoke grew thicker, they agreed that something more serious had occurred. Then the lights went out, and a dim emergency illumination automatically switched on. The smoke coming in turned dark brown and grew so heavy that visibility within the elevator cab became limited, and people began to cough and choke. Rinaldi pried the door open a crack and was faced with the close, solid wall of the elevator shaft. More smoke poured in through the opening. They had been trapped for about twenty minutes now, and their situation was suddenly desperate.

People reacted in different ways. The only woman crouched

silent in a corner as if she had gone into shock. One of the men grew hysterical—presumably because he was not an engineer. When I asked Rinaldi about it later, he said, "Yeah, we had a guy who was kind of emotional."

I said, "Crying?"

"Hyper and screaming and kind of upset."

The others ignored him, and he finally quieted down.

As for his own predicament, Rinaldi thought, "This is real." He was not afraid to die so much as he was worried about abandoning his wife and children. He locked eyes with an old friend named Frank Lombardi, a balding, soft-spoken man who at that time was the Port Authority's second-ranking engineer, and had his own family to leave behind. They exchanged little nods, acknowledging the danger they were in.

It was essential that they find fresh air, and soon. A lawyer named Rich Williams took off his undershirt and tore it into strips for people to tie around their noses and mouths. In desperation they pried the doors fully open. The wall that faced them was made of fire-resistant gypsum board two inches thick. Rinaldi, Lombardi, and Williams formed a chorus line and tried repeatedly to kick a hole through it—one, two, three!—but the wall merely flexed. Then Rinaldi remembered the drywall work he had done at home, and he suggested that they should score the surface. He took out his keys and immediately started scraping. Others crowded forward to join him. Using a key ring, one of them managed to unscrew a metal access plate that could serve as a chopping tool. The emergency lighting failed, and they continued to work under the green glow from a couple of electronic beepers. Eventually they dug all the way through, breaking a small hole into what seemed by feel to be an area of pipes. Fresh air came flooding in, diminishing the smoke inside.

They rested, and after a while returned to their habitual de-
liberations. A Port Authority engineer suggested that the fire had
been small and localized, and probably electrical. He was another one of those good company men with faith in procedures.
He said, "Someone will eventually find us."

In retrospect it seems significant that the counterargument,
which was essentially for improvisation, was made by Rinaldi. He
said, "No. Listen to the silence. There's not a sound in the building. There's no noise. Nothing. No one. You don't hear a thing.
No one is here in the building. This had to be larger than just a
small electrical fire. No one is going to come and get us. No one
is here. The building is evacuated, and they don't know where
we are."

After further discussion the engineers agreed that it made
sense to enlarge the hole and try to crawl through. The work
was done again in nearly total darkness, and it was slow. Rich
Williams took the lead. Finally he was able to reach across the
pipes, break through a thin wall, and get his hand into a new
space. He said, "I can feel this smooth, shiny surface."

One of the engineers said, "Maybe it's a bathroom. Does it
feel like tile?"

Williams said, "Yes. It feels like it could be tile." Then he said,
"Wait a minute. I have something else." He edged a little farther
forward, reached down to grab something, and slid back into the
elevator waving a disposable toilet-seat cover. It was a good moment. Nearly everyone laughed. As it turned out, they had dug
through the back wall of a toilet stall in the bathroom of the law
offices of the Brown & Root Corporation, on the fifty-eighth
floor. Williams was the first to go all the way through. "I made it
through," he said.

Rinaldi said, "Can you see light? Can you see a way out?"

Williams answered, "I don't know."

Rinaldi went through next, and dropped to the floor of the stall. He found the bathroom door and emerged into the deserted Brown & Root offices, where he went to a window to look outside. It was a gray winter day, not yet four o'clock in the afternoon, with a light snow falling. Far below, on West Street, stood a jam of emergency vehicles with flashing lights. There was no sign of fire. He returned to the bathroom, and he and Rich Williams helped the others as, one by one, they squirmed through the wall and, using the toilet as a step, escaped from captivity. After more than three hours in the elevator, all of them were filthy with soot and sweat, and coughing violently. They assembled in the office and began a long descent through pitch-black stairwells to the street. Like many of the survivors of 1993, they navigated in the darkness by linking hands, counting the steps, and talking themselves across the stairway landings. It was an uncertain and slow escape, nearly two hours long. But of course ultimately it succeeded.

Eight years later, on September 11, Rinaldi's memory of success informed his reactions during the first hour of the unfolding events, even after the second airplane struck and he knew that the Trade Center was under attack. His confidence was widely shared within the towers themselves, where people on the lower floors were isolated by the size of the structures from the horror overhead. Most knew less than Rinaldi did about what was happening, because of course they were not watching television. More than 15,000 walked out safely, in stairwells that were well lit and smoke-free. In 1993 the mood of the crowd inside the stairwells had been confused and fearful. This time there was little of that. Indeed, the mood was rather too restrained. In the South Tower, before it was struck, a Port Authority man made an announcement on the public-address system that the problem was contained to the North Tower, and that people could return to their offices—this despite the fact that jumpers could be seen falling into the plaza just outside. Most people ignored him. But in both towers a strange, Orwellian calm prevailed. People carried their morning coffee into the stairwells. They were not supposed to talk, and generally they did not. Those who tried to move fast or get ahead were reprimanded. But most were well behaved. They stayed obediently in lines to one side, leaving open lanes for the firemen to climb. Those who later believed that they had seen looks of knowing self-sacrifice in the firemen's eyes were understandably confusing the chronology of events. The looks in the firemen's eyes came from the extreme fatigue of having to hump heavy loads up so many stairs, as well as from the anxiety of moving into battle—in this case against a dangerous fire that would have to be fought from below. The firemen remembered the successes of 1993 as clearly as anyone, and they were no more prescient than others. But then, at 9:59 A.M., the South Tower made a noise variously heard as a roar, a growl, or distant thunder.

By September of 2001 Rinaldi's soft-spoken friend Frank Lombardi had risen to become the Port Authority's chief engineer—an exalted position with ties to monumental public projects of the twentieth century, not the least of which was the World Trade Center itself. Lombardi's office was grand. It occupied the southwest corner of the North Tower's seventy-second floor, and looked out over New York Harbor to the flats of Staten Island and New Jersey. As elsewhere in the towers, the view was restricted because the windows were only twenty-two inches wide and were inset one foot from the exterior—an unfortunate consequence of the close spacing of the external columns and also, apparently, of the discomfort that the principal architect, Minoru Yamasaki, felt with wide-open perspectives from such heights. Nonetheless, for a busy man like Lombardi the view was good enough. It was a backdrop. He didn't spend a lot of time looking outside.

In any case, the first airplane came in high and from the other direction—down the Hudson from the north, down Manhattan, and into the North Tower between the ninety-fourth and ninety-eighth floors. When it struck, it blew fuel and debris straight through the building. Lombardi was at his desk. He heard nothing, but felt the tower sway, and saw people in the hallway go airborne before they fell. His first thought was that New York was experiencing an earthquake. But then, just outside his window, he saw a huge ball of fire followed by a rain of vapor and white paper, and he realized that the building had been hit—though he assumed on the south face, and probably by accident. He went out to check on his staff. He told people to go to the stairwells and leave. Most of them did. He went back into his office and stared at the papers fluttering down, noticing how slowly some were falling, and how close to the window. From the quantity he realized that the perimeter wall had been badly breached. Right away he got three phone calls. The first was from his secretary's husband, who was in New Jersey, and had either seen or heard of trouble. Lombardi said, "She's fine. She's on her way down." The second was from the friend of a friend, who asked, "Is he okay?" Lombardi said, "Yes. He's on the way down." The third was from a contractor, who said, "Frank, do you need help?"

Lombardi wasn't sure. He said, "What happened?"

"I don't know, but something hit you."

"I know, but what?"

The contractor said, "I don't know."

It never crossed their minds to consider a hijacked airliner on a suicide flight. Lombardi wanted to verify that the seventy-second floor had been evacuated. With a couple of colleagues he searched the one acre of floor space, mostly cubicles. It was empty. But when he went to the stairwell to leave, he heard shouting behind an elevator door. Two men were trapped in the cab. Lombardi did not think they were in immediate danger, but having gone through his own elevator ordeal in 1993, he wasn't about to abandon them. The door was jammed, and required about twenty minutes to pry open. During that time one of Lombardi's colleagues was back in his office, getting tools, when the telephone rang. The man on the line said, "You guys are still up in Tower One? Don't you realize that Tower Two was hit?" In annoyance, Lombardi's colleague said, "Will you please stop spreading rumors?" and hung up. It seems surprising now. These were alert people, but they were confused by their proximity to an event too large to imagine.

Lombardi descended the stairwells of the North Tower to the plaza level, where he looked out and saw body parts scattered across the concrete. He went down another level, where all around him crowds were evacuating into West Street. But he was the chief engineer, and he felt a duty to respond—though how and to what he still had no idea. Prompted by memories of 1993, when a command post had been established in the complex's hotel (the Marriott, World Trade Center Three), he joined a few other Port Authority men and headed there through a passageway. They had assembled for a talk in the hotel bar along with some firemen when the place erupted in a tremendous roar. A pressure wave shattered glass, picked up the men, and threw them to the side. Lombardi thought that terrorists like those of 1993 had bombed the hotel and were maybe coming in through the doors, and he considered the irony that he had survived then only to die now, not 200 feet from where terrorists had hit before. The truth was stranger still: the South Tower had just collapsed over his head, and he had been saved by a few unusually heavy beams used in the structural splinting and patching up that he himself had directed after the earlier bombing. But he knew none of this at the time.

To his surprise, he felt nothing broken and no pain except for a burning in his eyes. That was widely the pattern of the day— survival as an all-or-nothing proposition. The room was dark, and so dusty that he had trouble breathing. He put a handkerchief to his mouth, which helped. Someone yelled to a fireman, "Could you please put on your flashlight?" The fireman did, to little avail. People stood up, saying, "Where are we? What's going on?" They lifted a steel roll-up door, thinking to get out, and found a group on the other side thinking to get in. The two groups frightened each other. A fireman went out to explore, and with visibility limited to about two feet, he nearly fell into a giant crater. He found a way across it and returned, saying, "Come on, I see a street-light." They went out in single file. Lombardi found himself on a sidewalk, but otherwise could see no change from the conditions that had existed inside. He lost track of his companions and walked down the street in confusion. He remembered the roar, and again thought of the bombing in 1993: had a device gone off in the underground? He passed the south pedestrian bridge, which he recognized, and headed south on West Street. He had a scratch on his forehead that was bleeding. People came up to him offering help, and someone gave him some water. Finally he got far enough away to look back. He saw the North Tower standing, but not the South. He thought, "Wait a minute. The North Tower is there. I *know* the North Tower is there. But what happened to the South?" It was confounding, and he could not conceive of an answer. He was an engineer, but human, too. He walked on for a while, until for the second time that day he heard a roar. He stopped and turned and watched in disbelief as the North Tower fell.

It took distance to sort things out. The technical speculation began right away—on the Internet, in engineering offices, and in universities around the world—and some of it was wild. Among the early embarrassments were statements by Van Romero, an explosives expert and a vice-president of the New Mexico Institute of Mining and Technology, who publicly opined that pre-set explosives could explain the collapses, and implied that the jet impacts were meant as bait for the rescue personnel. He was immediately inundated with e-mail from conspiracy buffs. After only a week he tried to take it all back, publicly renouncing his earlier ideas and signing on to the growing convention that fire and impact damage in combination were responsible for the collapses. Miserably, he said, "I'm very upset . . . I'm not trying to say anything did or didn't happen." But of course it was too late.

A Berkeley engineering professor by the name of Abolhassan Astaneh-Asl was more successful. He pulled down an early grant from the National Science Foundation, rushed to New York, checked into the trendy Tribeca Grand Hotel, got *The New York Times*'s ear, and began to show up on television inspecting twisted steel in the New Jersey scrapyards where it was being collected. Priscilla P. Nelson of the National Science Foundation helpfully explained, "It is important for engineers to be able to make the observations now that will lead to reduced building vulnerabilities in the future." A skeptic told me that some of the specimens Astaneh-Asl had identified on camera as bearing scars from the initial impacts were later discovered, from the numbering on them, to have originated much lower in the buildings. If true, it was hardly the sort of mistake that a man like Astaneh-Asl could have been expected to renounce, or even to mope about. This was a big disaster, with all kinds of impacts, and he had important work to do. In December he announced that he would be collaborating with a computer programmer from the Lawrence Livermore National Laboratory to build a "full-fledged, realistic model of the whole World Trade Center—including other buildings, not just the towers." He said, "Then we'll bring the plane in—to the computer model—and hit the building with various scenarios." Priscilla P. Nelson helpfully explained that the work "could lead to the generation of new computer models that will really enhance our ability to understand structural response." Astaneh-Asl may have revealed more than he meant to when he said, "The most important contribution of my career was to go to New York right after the attacks."

There was a lot of that kind of talk at the time.

But the main investigation had a different tone—cautious, preliminary, and publicity-shy. It was funded with $1 million from the Federal Emergency Management Agency (FEMA), and performed under the auspices of the venerable American Society of Civil Engineers (ASCE). The grant of such public authority to a private professional organization to investigate events potentially related to the competence of its own members would usually have raised legitimate questions about conflict of interest, but in this case it never quite did. Certainly there was an element here of a profession preparing to circle the wagons. The team was headed by an understated Midwesterner named W. Gene Corley, who holds a Ph.D. in engineering and is the senior vice-president of a modest research facility north of Chicago. Corley had led the official engineering review of the Oklahoma City bombing of 1995, and had an impeccable reputation as an independent investigator of building failures. He was a steady man, and as close to a disinterested insider as America was likely to produce. But he had limited control over a team that had expanded from a preliminary selection of about a dozen specialists—experts in metallurgy, fire, structural dynamics, and so forth—to an unwieldy group of twenty-two that included some of the profession's biggest names. The expansion was not just due to an urge to contribute, or to an abstract interest in the subject matter. People were also concerned that the investigation not provide ammunition for future lawsuits. The team was set up so that each of the members would be able to influence the contents of the final report. The pattern was exemplified by LERA, where the Trade Center's original engineer, Leslie E. Robertson, excused himself for obvious reasons of proximity—only to cede his place to his wife, Saw-Teen See, the firm's managing partner. But these were exceptional times. There was widespread agreement that for the health of the United States as well as the companies involved, it was important to keep the hyenas at bay.

Everyone understood the deal. The attack on the World Trade Center was an act of war. Despite the occasional chatter in the press about shoddy steel or substandard fireproofing, the towers were as well designed, built, and maintained as could have reasonably been expected in America in the late twentieth century. But the context had changed now. The towers fell because they were severely maimed and sprayed with burning jet fuel; they fell as any building will, no matter how resilient, if it is hit by the next bigger missile in the escalating progression of war. In retrospect, their greatest failing was so obvious that it hardly required discussion: the stairwells had been clustered too closely together, and their simultaneous destruction had trapped people above the impact zones, causing more than a thousand unnecessary deaths. This was an error that would have to be avoided in future designs. But there was no point in wishful thinking here. Civilians die in wars, and always will.

Beyond that, however, it was important as much in principle as in practice to make an honest effort to understand the process and mechanisms of each tower's collapse. This was to be a preliminary investigation only, with an emphasis on going public with the essentials as soon as possible. Corley knew that the effort was never going to be what enthusiasts in Washington were wishing for—something like an airline-accident investigation, controlled and contained, with the investigators on top, and the possibility even of a sexy reconstruction in a hangar as a finale. The Trade Center ruins were infinitely too large and chaotic to allow for that—and the emotions there, because of the continual discovery of bodies, were far too intense. The investigators spent some time at the site, but in practice their presence turned out to be of little help in the inquiry. Corley told me that given the weights and quantities involved, even the inspections of the steel that had been sorted at the New Jersey scrapyards proved to be an inefficient means of gathering evidence. This was certainly an argument against the critics who later claimed that all the steel should have been retained. More important were the videos and firsthand accounts from September 11, the original plans that LERA was providing, and the application of methodical science. Corley had confidence that the catastrophe could be reasoned through. It was obvious that critical inner workings of the collapses would remain obscure, and that debates and experiments concerning them would continue for years; indeed, this was an important part of the creative turbulence that followed the attacks, and that in time would lead to improvements in structural design. But as Corley had expected, by spring, as the last remains of the Trade Center were being removed, the essential facts were known.

The death of the South Tower was better understood than that of the North, because its symptoms showed particularly well. The 767 that approached from the south, United Flight 175, was a machine designed for lightness and efficiency, a slick aluminum tube 159 feet long with graceful wings stretching 156 feet from tip to tip, two fuel-efficient engines, and a cockpit that offered the pilots a gorgeous view. It was minimally loaded that morning for the Boston–Los Angeles run, with only sixty-five people aboard and about half of the maximum fuel, and as it approached the building it weighed 137 tons. It was flying at nearly 590 mph, which was more than 150 mph above the airplane's designed limit at low altitude. In the cockpit the overspeed warning must have been warbling loudly.

The tower that soared above the cockpit in those last seconds was in its own way designed for lightness and efficiency—and it, too, was essentially a tube, though squared, 1,362 feet tall, and made of steel. Each of its faces was 209 feet wide and consisted of a palisade of aluminum-clad fourteen-inch box columns, which when bolted and welded together and braced by horizontal plates formed the exoskeleton that carried half the building's weight, and gave it all of its stiffness against side loads. The side loads were expected to come primarily from winds and, at the design limit of 150 mph, to amount to a maximum sustained 13-million-pound push. But the building was more resistant even than that, because the Port Authority, out of conservative impulse, had decided during the construction to go beyond the design and anchor the foundation structure directly into bedrock. Hidden inside the square tube of perimeter columns stood the other half of the support structure: a central core consisting of massive columns arranged in a rectangle much longer than wide, which was aligned in the South Tower on a north-south axis and contained the elevators and emergency stairwells. Though the core was not designed to resist side loads, its alignment meant that the north-south axis was, however marginally, the tower's strong one. More significant, it also meant that the acre of column-free office space on each floor (the goal of this tube-inside-a-tube design) was wider on the east and west sides (think viewless cubicles) and narrower on the north and south. The floors were built on thin steel trusses that bridged between the core and perimeter columns and supported corrugated-metal decks on which concrete was poured. The floors were thin and light, and served as diaphragms to stiffen the exterior columns against buckling forces. Because the vertical load lessened significantly with height, the thickness of the steel in the columns was progressively reduced from the bottom of the tower to the top. But no matter how it was trimmed, the weight added up. Ignoring all the people, furniture, and equipment that the tower contained, its structure alone weighed some 600,000 tons.

By comparison, the oncoming 767 was a mere pebble. But as pilots know, the energy of an impact varies with the square of the airplane's speed—meaning if you go a little faster, you hit a whole lot harder. And this airplane was moving ferociously fast. The killer at the controls almost missed, and would merely have clipped the left wing, but he hooked his intent last turn like a wolf after prey, and punched hard into the south face of the tower three fourths of the way up, on the east side of center. The airplane hit about nine times harder than it would have during a typical accident while on an approach to landing. The building never came close to toppling sideways, but it swayed as much as twelve inches at the top, and endured a violent impulse that ran vertically through it like a rebounding wave. The airplane entered the building with its wings banked left, stretching from the seventy-eighth to the eighty-fourth floor, slicing through half the south-face perimeter columns. Frame by frame, the videos show it heading inside as if intact. An MIT study later estimated that only 6 percent of the plane's kinetic energy was expended against the south face. Almost all of the rest of the energy was expended inside—25 percent in tearing up the floors and pushing and piling the cubicles, and another 56 percent in damaging or severing as many as half the columns in the core. Two of the three stairwells were taken out, along with the people who were using them to evacuate the building in response to the earlier North Tower attack. The third stairwell was damaged but remained usable, and allowed several brave and desperate people trapped above the impact zone to escape. A few of the airplane's heaviest parts—an engine, a piece of landing gear, a wheel—kept right on going, as far as six blocks into the city, and a section of fuselage hit the roof of World Trade Center Five. But most of the airplane remained inside the intended victim, piled into a dike of mixed debris against the tower's north and east perimeter walls.

One of the many astonishments of that day is that the building was able to swallow an entire 767, and to slow it from 590 mph to a stop in merely 209 feet. According to calculations based on photographic evidence of the damaged perimeter, the structure had redistributed the gravity loads so efficiently that only twenty feet from the entry wound the demands being made on intact columns were hardly higher than normal. The building was cer-

tainly not safe. It had suffered serious if unknowable damage
to the east side of the core. At the same time, for seventy feet
along the inside of the south face, and perhaps a short distance
along the east face, the floors had been torn loose from the structure, leaving perimeter columns unbraced under the gravity load
of the roughly thirty-story building overhead. Still, if not exactly
shrugging off the hit, the South Tower had absorbed it well: absent an earthquake or a strong windstorm, the building would
have remained standing indefinitely.

But then, of course, there was the fire. As the airplane disintegrated, it released its fuel—roughly 10,000 gallons of volatile
"Jet A" kerosene that sprayed and vaporized through the six
floors of wreckage, poured into the elevator shafts, shot through
broken windows along the north and east faces, and immediately
ignited. Apocalyptic though it seemed, the huge fireball that
blossomed over the plaza was actually to the building's advantage, because it consumed as much as a third of the available
fuel, releasing the heat harmlessly into the air. That left two
thirds of the fuel inside the tower, however, and it was widely
spread and burning.

The building's sprinkler system had been destroyed by the
impact, but in any case it would have lacked the pressure to operate effectively over such a wide area. More significant, much of
the light spray-on fireproofing that coated the structural steel had
been knocked away—and precisely in the same east-side areas
that had sustained the worst damage and now were threatened
by unusual and increased loads. On only one of these floors had a
fireproofing upgrade that doubled the thickness of the coating
been completed—a circumstance that provided for healthy debates afterward, though there was never direct evidence that the
thicker fireproofing would have better endured the impact.

At any thickness, the coating was meant to insulate the steel and prevent it from overheating in a fire for at least two hours (by which time an office fire will typically have burned itself out in any given part of a building). The problem with overheating is that long before steel melts, it weakens. In structural steel like that of the South Tower the weakening begins around 350°. By 1,100° the steel loses about half its strength. Kerosene burns twice that hot inside the carefully crafted furnace of an airliner's engine, but within the imperfect combustion chambers of the South Tower impact zone it is thought to have reached at most only about 1,500°—a temperature unlikely to induce the failure of such a tremendously redundant design, especially on a gentle, windless day. Moreover, the kerosene fire simply did not last long enough at any temperature to overheat the building's massive columns. Corley's fire specialists believed that the jet fuel never collected into deep, slow-burning pools, and that it burned through entirely within four minutes. They concluded that jet fuel itself did not bring the tower down.

What it did do, however, was set off raging office fires simultaneously on six different floors. It was a conflagration that would have been impossible for the firemen to control, had they gotten to it—a fire large enough to create its own powerful winds, sucking oxygen in through all the perimeter holes and broken windows, generating energy three to five times greater than that of a standard nuclear power plant, and eventually heating the steel to temperatures as high as 2,000°. The fire fed on wrecked office furniture, computers, carpets, and aircraft cargo, but primarily it fed on ordinary paper—an ample supply of the white sheets that were so much a part of the larger battlefield scene. Without that paper, Corley's experts believed, the fire might not have achieved the intensity necessary to weaken the steel beyond its critical threshold. It would be simplifying things, but not by much, to conclude that it was paperwork that brought the South Tower down.

On the debris pile in the northeast corner the fire melted the remnants of the shattered airliner, which half an hour after entering the building began to flow in a stream of molten aluminum down the tower's outside. Still the tower endured. But then the southeast corner of the eightieth floor collapsed, triggering the progressive failure of the floor along the entire east side of the building. Dust plumes shot out through the broken windows. The eastern perimeter columns now stood unbraced for the space of at least two floors, adding to the dangers that the building already faced from the unbraced condition on the perimeter on the south side. Shortly afterward, in the southeast corner, near the entry wound, a cluster of exterior columns began to buckle. It was 9:59 exactly, merely fifty-six minutes after the airplane hit. Peter Rinaldi was living the last four seconds of his earlier life. Frank Lombardi was pushing into the hotel bar somewhere far below. With its support giving way beneath it, the top of the tower tilted east and then south, rotating in a clockwise direction, and suddenly slammed down. Even if it had been felled from below, the tower could not have capsized in a conventional sense, because like most other buildings, it lacked the structure to hang together for more than a few degrees off vertical. But it was not felled from below; it was hammered from above, and it accelerated as it fell, crushing the core and peeling back the exoskeleton with each successive floor. As the external walls peeled, they broke primarily at the bolted connections, allowing welded prefabricated sections of the columns to fall free. The upper sections fell east and south, and hit the Bankers Trust building; the lower sections fell north and west, and gave the Marriott hotel the first of its two fatal blows.

The North Tower died a half hour after its twin, but because it showed fewer symptoms, less is known about its end. The attack it suffered was similar. It came from another 767, American Flight 11, which was light on passengers and fuel, and flying 100 mph slower than the United flight, though still very fast. The initial impact involved a full, clean entry, almost perfectly centered. The airplane sliced through thirty-six of sixty-one columns on the north face, tore through four floors, and slammed squarely against the weak axis of the core. It took out all three stairwells, destroyed the sprinkler system, knocked the fireproofing from the steel, and blew a piece of landing gear and a fireball through the far wall. Lifejackets and parts of the seats ended up on the roof of the Bankers Trust building. The North Tower swayed, the jet fuel was rapidly consumed, and a terrible office fire broke out.

There were differences, too. The North Tower had the advantage of getting hit high up, thereby requiring less performance from the weight-bearing steel in the fire zone. This must certainly have saved lives. The external columns held. But by 10:28 their integrity was not enough. In this design of a tube within a tube, both tubes had to stand for either to survive. The North Tower core was aligned on an east-west axis, and it had been severely damaged—probably along its full length—by the airplane's centered impact. After nearly two hours of progressive weakening by fire, the remaining columns reached their limits. There was no sign of this on the outside. The South Tower of course had already fallen, and smoke was rising from a wound near the North Tower's top. Nonetheless the building remained a monolith, seemingly as permanent and strong as stone. But then the 351-foot transmission tower on the roof sank a little. The movement was barely perceptible. Half a second later the floors above the impact zone dropped as a unit straight down through the office fire, creating a flare-up and the illusion of a secondary explosion before striking the first blow in the chain of blows that pancaked the monolith to the ground.

Is there any wonder that Frank Lombardi, the Port Authority's exalted chief engineer, had no understanding of what he had just seen? For the second time in an hour the Trade Center's dust darkened the sky. Lombardi walked to the Hudson's edge, where the dust cloud abruptly ended, and after a while he found his way onto a small ferryboat—one of several that had responded to the disaster and were nosing up to the shore, evacuating the shell-shocked survivors. The boat took on a full load of them, and headed across the river for New Jersey. Lombardi stood in the crowd near the stern, and gazed numbly at the receding city. From a hole where the Twin Towers should have been, a column of smoke and dust rose into the blue of the sky and, widening slowly, streamed out toward Brooklyn. This was the view already made famous around the world by television, evoking widespread fear and anger, and in some places celebration. Encountering it in reality on this river, with a warm wind on his face and the sound of the ferry's engines in his ears, Lombardi found it more than strange. By then he had heard a rumor about the cause—two airliners, it was said, involved in a coordinated attack—but people everywhere were full of frightened chatter, and he had no way to judge it. He may have sensed that he of all people should have been able to reason things through, but he could not. This must have been deeply disturbing to a man who was nothing if not pragmatic.

The ferry kept moving confidently toward New Jersey; that much at least was normal. But then, as it approached the far shore, its engines suddenly idled, and the captain leaned out of the pilot house and called, "Man overboard!" Lombardi checked the wake for the victim, and to his continuing confusion saw no one there. Instead, someone was in the water ahead—and he appeared to be swimming toward the Trade Center site. The captain brought the boat slowly up beside him, and the crew dropped a ladder. The swimmer was tiring. He reached up, and was about to be helped aboard, when someone shouted, "Don't let him in! Shoot him. He may be a terrorist!"

This halted the proceedings.

A policeman stepped forward uncertainly, and leaned over the rail. He said, "Excuse me, but what are you doing in the water?"

The swimmer said, "I thought I could swim over to New York to help people."

The policeman hauled him up on deck. He said, "Let me get this straight. Let me understand this. We're bringing people over on a boat from there to here. Right?"

"Yeah."

"And *you* want to swim over there?"

"Right."

The policeman did what policemen do, maybe for lack of imagination. He said, "You're under arrest." He probably figured the man was unbalanced, and dangerous at least to himself. He was still holding him when Lombardi got off the boat.

Lombardi found his way to the Port Authority offices in Jersey City, above a commuter station known as Journal Square. It was there in the early afternoon, when he saw the events replayed on television, that he first was able to understand what had happened to the buildings, and to begin the slow process of acceptance. Two months later, when I met him in midtown Manhattan, he had come a long way but was still not all the way there. He had a manner of talking about the towers as if they had once been alive. He had obviously been thinking heavily about their deaths.

He knew many details of the attacks. He told me he thought that the terrorist who came in first, against the North Tower (in which Lombardi himself was sitting), had intended to knock it right over—a motive that explained the perfect centering and the high hit against the weak axis of the core. Furthermore, he proposed that the second terrorist, while crossing the Hudson and setting up his own attack, had observed the wounded building still standing, and at the last moment had decided on a strategy of avoiding a direct hit against what he knew was the strong axis of the South Tower's core, instead choosing to hit the building low and on one side, in an attempt to undercut it and trigger a collapse. If so, he was fantastically nimble. I had another view of the attack—that the terrorists were operating not as engineers but as executioners, that they were imbued with clumsy metaphors of swift swords and decapitation blows, and that the second one had simply missed his stroke by a few feet, as is easy to do with a fast-moving airplane. Lombardi, I thought, was sometimes still thinking like a defeated man, investing too much competence in his enemies.

He was a gentle person. You could see him actively trying to recover his equilibrium—swinging between the old-fashioned Port Authority can-do confidence, which was coming back, and a persistent fatigue, a vagueness that was said by his friends to be something new. In many ways he was the ultimate Port Authority man, and he seemed to embody the organization especially in its ambivalent response to the Trade Center disaster. He rarely came to the site. It was as if the loss there was still too painful to contemplate. But he also told me that he was proud and glad for Peter Rinaldi, his chosen representative, who had seized this chance to take action.

Rinaldi was the lucky one. He worked seven days a week, often fifteen hours a day. His base was the city's emergency operations center, in the kindergarten rooms of Public School 89. The school occupied the second floor of a new red-brick building just inside the emergency perimeter, two blocks north of the Trade Center site on West Street. It had been safely evacuated on September 11 before the towers fell. When the city's recovery team set up headquarters there a few days later, there was time only to dump the food from the children's lunch boxes so that it would not rot. The lunch boxes themselves stayed in the classrooms, piled in rows on the shelves, with the names still displayed: Patrick, Elizabeth, various Jennifers, and all the other children who had fled with their parents to safe havens and new schools elsewhere in the city or beyond. For two months afterward the classrooms remained as they were, with strip posters of the alphabet taped to the chalkboards, and colorful, childish art pinned to the walls.

The inner sanctum was a corner room with cinder-block walls painted creamy white, and linoleum floors now brown with the dirt tracked in from the site. The shelves that lined the walls were still filled with blocks, toys, and plastic crates of Dr. Seuss and other children's books. At the center stood four folding tables that had been pushed together, along with some folding chairs that had been scavenged from somewhere. When the tables grew crowded, people simply used the kindergarten stuff—pint-sized furniture built low to the ground, which was discovered to be strong enough even for 250-pound construction workers. Visitors often remarked on the strangeness of the scene, in which full-grown men and women, including powerful executives and renowned engineers, sat around as if they were playing at being children again. In fact the stakes were large, and so were many of the egos involved. But the room did not allow for displays. People there dressed alike in dirty boots and rough clothes, and no matter who they were, they had to prove themselves again. Their conversations were held in the open for lack of choice. There was no time for memos, or for chain consultations. The e-mail connection was permanently down. The phones did not work. When problems arose, they were dealt with right away, either in the room or, if more information was needed, with a walk down the street to the pile, and a decision on the spot. This is what Peter Rinaldi encountered when he first arrived. As the operational center of a billion-dollar effort, Public School 89 was a highly unusual place.

Leading the effort was the unlikely duo of Kenneth Holden and his lieutenant, Michael Burton—the two Department of Design and Construction officials who had emerged from bureaucratic obscurity on September 11 to orchestrate an effective response to the disaster. Holden, the DDC's shrewd and intellectually sophisticated commissioner, was a heavyset man who had grown up in an atmosphere of suburban ease, had worked on a kibbutz in Israel, and had studied history at Columbia. Having drifted by chance into a city-government job, he was now, at forty-four, one of the youngest department heads in New York City government. Burton was a slightly younger man, a thirty-nine-year-old marathon runner and construction-industry insider who, having been raised in modest circumstances, had earned a degree in engineering and an M.B.A. from colleges in the Bronx, and was aggressively climbing the ladder of social and material success.

The two men made an awkward pair. Each had rushed independently to City Hall at the first news of the attack, and had been caught on the streets and forced to run when the towers collapsed. By that afternoon they had met up and, with no thought beyond a few hours into the future, had begun to enlist the help of others within the DDC, and to organize the arrival of expert engineers and heavy equipment at the Trade Center site. They did not ask permission to do this—nor, at first, did anyone pay much attention to their work. The DDC was not even mentioned in the city's official emergency-response plan. The focus over the first few days was mostly on the drama of the bucket brigades. But as is widely known, Mayor Giuliani had smart reactions to the crisis. Later it seemed that one of the smartest was a back-room decision to scrap the organization charts, to finesse the city's own Office of Emergency Management (OEM), and to allow the DDC to proceed. The federal government was poised to intervene, but agreed to hold off, and then to hold off again. The result was the classroom at PS 89, that strange kindergarten scene through which control of the contractors was exercised and federal funds flowed.

Holden spent months on his cell phone there, marshaling resources for the operation, defending it from political attack, and creating the shelter in which Mike Burton then could operate. Burton's job was to oversee the practical details of the cleanup. He was tireless: he excused himself from his home life in suburban Westchester County, moved to an empty apartment in Battery Park City, and began spending eighteen hours a day at the site, roaming the pile even late at night, searching for efficiencies, forever urging the crews to work faster. He was not a popular man. He was extremely ambitious. He was determined to finish the project in less than a year. The firemen in particular distrusted him for what they saw as his disrespect for their dead. There were days when it became awkward for him to go out onto the pile, because of his sour relations with the men there. Not without irony, people called him the Trade Center Czar. But such a boss was probably needed to get the job done. The scale and complexity of the ruins required it. And on a level just below him there were too many other bosses who did little more than eat and mill around.

Indeed, for the first two months the site's outer zone was so dense with agencies and their equipment that the debris trucks at times could not get through, and the hauling operation had to shut down. It sometimes seemed that every official with a uniform or a badge within a hundred miles wanted to put in time there. Even the dogcatchers from Long Island showed up. They hauled in a mobile spay-and-neuter clinic to use as a command post, and sat around on lawn chairs for weeks, wearing T-shirts imprinted in block letters with SPCA—LAW ENFORCEMENT. They carried pistols, too—part of the impressive brandishing of weapons all around the ruins. One afternoon, as we walked toward the pile, Pablo Lopez, the indomitable Ecuadorian engineer, said to me, "When you see even guys from the EPA carrying guns, you've got to wonder. What's an environmentalist doing carrying around a Glock? Who's he gonna shoot?"

The weaponry made people seem impotent and out of touch, as if they had showed up too late for the fight. The frustration was that you couldn't dislodge the debris by shooting it. Because of the bodies that lay there, you couldn't dynamite it either. Because of New York's sensitivities to noise and dust, you couldn't even use small demolition charges to fill dangerous cavities or bring down the skeletal walls. There was no choice but to cut and pull and unbuild the chaos one piece at a time. Burton understood this early. While others stood around with their guns, he and Holden imposed a rough order on the site by dividing it into quadrants, roughly equal zones that he assigned to four large construction companies: Turner, AMEC, Bovis, and Tully. The first three were essentially management teams—multinational corporations in the high-rise construction business, which had little equipment of their own and relied on other contractors to get the work done. The fourth company, Tully, was quite a different thing—a family-owned New York paving contractor with little structural experience but plenty of trucks, heavy equipment, and experienced workers of its own. All four companies had teamed up with the DDC before and were major players in New York.

The quadrants were never as neatly delineated on the ground as on the site maps, but during the critical opening stages of the project they served to concentrate each company on the debris immediately at hand, and to keep people from being overwhelmed by the magnitude of the process that lay ahead. The effect across time was a shaping of the pile according to the respective corporate personalities: Turner in isolation on the ruins of World Trade Center Seven, efficiently disassembling the debris and making almost tidy work of it; AMEC in the northwest corner of the foundation hole, often bumbling, arousing the ire of the DDC, and falling behind; Bovis in the southwest corner, progressing steadily in the difficult area of the Marriott ruins and the south skeletal wall, while playing off its management skills to spread incrementally throughout the site, most significantly by taking on the all-important job of saving the foundation hole's walls; and, finally, Tully, everyone's favorite, the hardworking red-meat guys, who kept pulling recklessly ahead and devouring the ruins along the entire east side.

These were the forces, roughly 3,000 strong, that Ken Holden and Mike Burton unleashed. In a painful and probably necessary process they slowly wrested control from the firemen. In late October the operation became a "joint command" between all the uniformed services and the DDC. Realistically, because the uniformed services were split by long-standing rivalries and had little of technical value to contribute, the joint command left the civilians in charge—a fact that became increasingly obvious at the regularly scheduled "unified" meetings, where the Fire Department retained a functional veto power, but only the DDC seemed capable of moving ahead.

Emotions were raw. One of the unacknowledged aspects of the tragedy was the jealous sense of ownership that it brought about—an unexpected but widespread feeling of something like pride, that "this is our disaster more than yours." The feeling started at large in the United States, and became more acute with proximity to the site—a progression of escalating possessiveness that ran from the halls of Washington through suburban New Jersey to New York, and from there through Lower Manhattan to the pile itself, where it divided the three main groups (fire, police, and construction) and sometimes set them against one another. The firemen in particular felt that they had a special relationship with the site, not only because they had lost 343 people there—out of a force of 14,000—but also because afterward their survivors, along with their dead, had been idolized as national heroes, and subjected to the full force of modern publicity. A few of them reacted embarrassingly, by grandstanding on television and at public events, striking tragic poses and playing themselves up. Even at the site, where people generally disliked such behavior, you could find firemen signing autographs at the perimeter gates or, after the public viewing stand was built, drifting over to work the crowds. Most of them behaved more soberly, no matter what pride they may secretly have felt, if for no other reason than that their firehouse culture had until now frowned on self-aggrandizement. Still, there was resentment by the police, who had lost plenty of their own people, and by the construction crews, who took it upon themselves to remember the far greater number of civilian dead. These tensions flared especially over the differing treatment of human remains—on the one extreme, the elaborate flag-draped ceremonials that the firemen accorded their own dead, and on the other, the jaded "bag 'em and tag 'em" approach that they took to civilians. A strange blindness caused them to persist with this behavior despite the ease with which it could have been remedied. Even Sam Melisi participated in it, for instance once bemoaning a "drought" to me when the remains being uncovered were merely those of civilians. It was a surprisingly ganglike view, and it encouraged a gang mentality among others on the pile.

Still, most people at the site understood that the loss the firemen felt was real. One morning in the late fall, I accompanied Ken Holden into the expanding valley at the center of the pile, where a temporary access road of ground-asphalt millings was being built. A fire chief came up and said, "You gotta give us time. You gotta get these guys to stop covering up the debris, burying us with dirt."

Before Holden could answer, another fireman, an older man in filthy clothes and a scarred helmet, rushed up and said, "You stop these guys from pushing dirt in here!" He had a weathered face, and heavy sweat on his upper lip. His eyes were wild. He said, "I've got two friends out there. And I've got my son buried right in here." Holden put a calming hand on his shoulder, to no effect. The fireman wandered off with his shovel, a short entrenching tool bent 90 degrees, and climbed down into a hole in the rubble.

The chief repeated almost apologetically, "He lost his son in there."

Holden said, "How much time do you need? A day? Two days?"

The chief swept his hand wide. "They're buried all through."

Holden went off to find Burton and tell him to stop the road for now. When he came back, the old fireman was down on his knees, probing loose debris and sniffing shovelfuls for the scent of death.

For most of the men on the pile, the loss was less acute. Even if these were fellow firemen who had died, and even if they were close friends and even if people used the word "brother" to describe them, it was not the same as losing a son. Still, over the first few months there was a lot of sadness at the site. Away from the photographers and TV cameras, the depth of it was not always obvious. The Fire Department search parties operated on a regularized schedule in small groups beside the diesel excavators, and they sifted through the fresh debris with workmanlike efficiency. But they also took risks for no obvious reasons—jumping suddenly into newly opened debris holes, climbing on the unstable cliffs, and, especially, standing for hours in the heaviest smoke and dust, refusing as a matter of pride to wear the respirators that dangled around their necks. They seemed to have surrendered to an attitude of reckless self-abandonment. To varying degrees the police and construction workers had surrendered to it too. In one of the Salvation Army feeding tents I talked to a psychologist who blamed the risk-taking on "survivor guilt," which he called a common reaction to disaster. To me, however, it looked like a simpler form of grief.

But the risk-taking was also the expression of a more creative and courageous impulse, linked to the need for action and improvisation, and part of the personal freedom that was unexpectedly emerging at the Trade Center site. It was a response as well to the sheer magnitude of this calamity, the hijackings, the killings, the collapses—events so extreme that they required extreme actions in return. Despite all the divisions at the site, this was the fundamental understanding that people shared—and, indeed, that the culture at the site demanded.

And then, of course, there was the pile, always the pile. It had been the focus of ferocious energy during the collapse, and now again was the focus during the unbuilding. The pile was an extreme in itself. It was not just the ruins of seven big buildings but a terrain of tangled steel on an unimaginable scale, with mountainous slopes breathing smoke and flame, roamed by diesel dinosaurs and filled with the human dead. The pile heaved and groaned and constantly changed, and was capable at any moment of killing again. People did not merely work to clear it out but went there day and night to fling themselves against it. The pile was the enemy, the objective, the obsession, the hard-won ground.

THE RUSH TO RECOVER

The dread that Americans felt during the weeks following the September 11 attacks stemmed less from the fear of death than from a collective loss of control—a sense of being dragged headlong into an apocalyptic future for which society seemed unprepared. That future was first delivered at a precise place and time—inside American Airlines Flight 11, a twin-aisle Boeing 767 that had departed Boston and was westbound for Los Angeles at 29,000 feet, near Albany, New York. This was the airplane that would hit the World Trade Center's North Tower. Both pilots were at their accustomed positions in the cockpit. At 8:13:29 the en route air-traffic controller at Boston Center said, "American 11, turn twenty degrees right." One of the pilots answered, "Twenty right, American 11." It was the flight's last routine transmission. Sixteen seconds later the controller issued instructions for a climb, and received no response, presumably because in that brief interval the hijackers had burst into the cockpit and with brutal efficiency had assumed command.

The controller did not suspect it at the time. Indeed, neither he nor others on the ground could have guessed that anything about the morning was extraordinary. In New York City, Sam Melisi was sitting in the familiar engine room of the fireboat where he worked, studying for a ship's engineer exam that he was scheduled to take at the end of the week. Ken Holden was in his Queens office preparing to leave for a regular meeting in Lower Manhattan, at City Hall. Mike Burton was headed to the same meeting, and was weaving impatiently as usual through traffic on the drive from his house in suburban Westchester County. At the same time, tens of thousands of office workers were streaming from the subways and commuter trains, and grabbing their customary coffees, or riding the elevators straight up to their offices in the World Trade Center towers. Workers all through the city were doing similar things. There were perhaps 18 million people in the vicinity of New York that morning, and later it seemed every one had a story that started with a routine.

The routine held at Boston Center for eleven long minutes after the hijacking began. The weather was known to be bright, smooth, and clear. Air traffic was light. For the controller who had lost radio contact with American 11 there really wasn't much to do. The airplane tracked steadily westward across the radar screen. The controller assumed that its continuing silence amounted to a simple communication failure—a relatively common occurrence—and he tried to re-establish radio contact, but without undue concern. He mentioned the problem to another controller, in charge of the adjoining airspace, and he continued to work the other flights in his piece of the New England sky— a Continental, a couple of Deltas, and two business jets that cruised by.

Then came the next step in the sequence of what would become a strange aerial ballet: at 8:23 the second 767 destined to hit the Trade Center came flying into the controller's sector and as expected checked onto the frequency. It was United Flight 175, running fifteen minutes behind American 11, and also bound from Boston to Los Angeles. This was the same airplane that would soon flow in a stream of molten aluminum down the South Tower's side, but there was no hint of that now. The pilot sounded particularly relaxed. He said, "Boston, morning. United 175 out of nineteen for two-three-zero."

With a confirming glance at the altitude readout on the radar screen, the controller said, "United 175, Boston, ah, Center, roger." The frequency was silent for nearly a minute and a half. Then the controller heard something in his headset, possibly just the hollowness of an empty "carrier" signal—a raw transmission without the modulation of spoken words. He radioed, "Is that American 11 trying to call?"

This time there was a voice, heavily accented. On the public air-traffic control frequency, for all who were tuned in to hear, it said, "We have some planes. Just stay quiet, and you'll be okay. We are returning to the airport."

Little is known about the scene in American 11's cockpit at that point, since the airplane's cockpit voice recorder was never found. No "black box" recorder survived in the Trade Center's ruins. It is possible, as was reported afterward, that one of the pilots had surreptitiously keyed his microphone to pick up the threats and warn air-traffic control. But given the design of aviation microphones, which require close proximity to the mouth, it seems more likely that the pilots were already dead or disabled, and that one of the hijackers had strapped himself into the seat behind the controls and was misusing an audio selection switch, unintentionally transmitting a message meant only for the cabin's public-address system.

It had been fourteen years since the last airline hijacking in the United States. Understandably, the controller was taken aback. He radioed, "And, uh, who's trying to call me here?"

Silence.

He radioed, "American 11, are you trying to call?"

Again the hijacker was heard. He said, "Nobody move. Everything will be okay. If you try to make any moves, you'll endanger yourself and the airplane. Just stay quiet."

The controller caught on, and did not radio American 11 again. He notified his superiors that a hijacking was in progress. The United pilots must have overheard all this and understood it too. Like the other crews on the frequency, they maintained radio discipline. Not a word of comment went out over the air. Two minutes later the controller handed off the United flight to another sector within Boston Center's airspace.

It was 8:28. American 11's transponder beacon had been switched off, degrading its display on the radar screens and extinguishing its altitude readout. As an unenhanced "primary" blip, it turned sharply and began to move at high speed down the Hudson Valley toward metropolitan New York. As it busted through the various sectors and into the airspace belonging to New York Center, the Boston controllers continued to monitor the last frequency—now cleared of other traffic. The flight downriver lasted eighteen minutes. A third of the way through it the frequency came alive with another unintentional transmission: "Nobody move, please. We are going back to the airport. Don't try to make any stupid moves." It sounded almost polite. People took the words at face value. The destination appeared to be New York. Newark and Kennedy Airports prepared for an emergency landing.

Then United 175 got involved again. The pilots had checked onto a new Boston Center frequency, and were level at 31,000 feet, when the hijacked airplane flew nearby. The controller wanted to verify its altitude, and he asked United 175 for help. He said, "Do you have traffic? Look at, ah, your twelve to one o'clock position at about, uh, ten miles southbound, to see if you can see an American 767 out there, please."

Having heard the transmissions on the previous frequency, the United pilots must have known what this was about, but again they maintained the expected calm. The view from their cockpit was superb. Looking downsun, they spotted the airplane ahead and several thousand feet below, crossing fast from right to left. "Affirmative. We have him, ah, he looks, ah, about twenty, yeah, about twenty-nine, twenty-eight thousand."

Reflexively the controller directed them into the airspace above and behind American 11's tail. He said, "United 175, turn five . . . turn thirty degrees right. I want to keep you away from this traffic." The encounter was later reported as a near collision between the two airliners—a gross if typical exaggeration. Ultimately, of course, the turn did neither airplane any good.

The scene inside American 11 was very rough. There were five hijackers aboard, Islamic militants armed with box cutters and small blades. They were led by the now notorious Egyptian named Mohamed Atta, the chief conspirator behind all four of the attacks that day. The cabin of American 11 was less than half full, with seventy-six hapless passengers and nine flight attendants. One of the flight attendants, a forty-five-year-old woman named Betty Ong, dialed a seatback flight phone and reached a reservation agent on the ground. In terrified tones, gasping for air, Ong reported the hijacking. The agent passed her to a supervisor, who patched her through to American Airlines' national operations center, in Fort Worth, Texas. The manager on duty there, an airline veteran named Craig Marquis, pulled up her records and, concerned that the call might be a hoax, asked for her employee number and nickname. This she managed to give him. She said that two of the flight attendants had been stabbed, one so severely that she was on oxygen, and that a business-class passenger had been killed by having his throat cut. She counted four of the hijackers, and reported their seat numbers. She said they had used a chemical spray that burned her eyes and made breathing difficult. Marquis could do nothing but keep her company. He asked Ong if there was a doctor on board. "No, no doctor," she said. As the airplane approached New York, he asked her if it was descending. She said, "We're starting to descend! We're starting to descend!" By then she may have felt more hope than horror. It was natural to assume that they were descending to land.

Apparently Ong could not see into the cockpit, which was just as well if the pilots inside it had been killed. Mohamed Atta was at the controls. Descending along the Hudson River, he pushed American 11 to over 500 mph, nearly twice the normal low-altitude speed, and hardly what you would expect from a 767 setting up for a landing. Still, no one yet guessed his purpose. At 8:40 he was six minutes out from the North Tower.

The pilots of United 175 had lost sight of him in the vastness behind their left wing. They checked onto a new frequency, now in New York Center's jurisdiction, with an abbreviated call: "United 175, at Flight Level 310."

The controller answered, "United 175, roger," but his mind, too, was on the airspace behind them. He radioed to another crew who had been asked, like the United pilots, to spot American 11 to estimate its altitude. "USAir 583, do me a favor. Were you asked to look for an aircraft, an American flight about eight or nine o'clock, ten miles, southbound, last altitude two-nine-zero? No one is sure where he is."

USAir answered, "Yeah, we talked about him on the last frequency. We spotted him when he was at our three o'clock position. He did appear to us to be at twenty-nine thousand feet. We're not picking him up on TCAS."

The controller said, "No. It looks like they shut off their transponder. That's why the question about it." This was by then so obviously an inadequate explanation that the controller seemed to be talking in code—signaling that something very serious was occurring here. The frequency was quiet for a full minute, during which the United pilots apparently decided to drop any pretense that American 11's problem was merely a communication failure. Even so, when they radioed again their expression was tightly controlled. "New York, United 175 Heavy."

"United 175, go ahead."

"We figured we'd wait to go to your center. We heard a suspicious transmission on our departure from Boston. Sounds like someone keyed the mike and said, 'Everyone stay in your seats.' " He was referring, of course, to American 11's unintentional transmissions.

The controller said, "Okay, I'll pass that along."

The pilot said, "It cut out."

That was United 175's final call. The time was 8:42. Just afterward a team led by a twenty-three-year-old citizen of the United Arab Emirates named Marwan al-Shehhi invaded the cockpit. It is obvious that, as in American 11, they busted through the door. The attack was so sudden that the pilots had no chance to alert the world. Nonetheless these 767s were big things with hidden spaces and lots of telephones, and word quickly got out: a mechanic on duty at United's San Francisco center for in-flight complaints got a call from a flight attendant on board who, before the line went dead, blurted, "Oh, my God! The crew's been killed, and a flight attendant has been stabbed! We've been hijacked!"

The airplane held five hijackers, two dead pilots, seven flight attendants, and fifty-one passengers. United 175 turned and dove toward New York.

At about the same time, American 11 passed low over the George Washington Bridge. As best the last moments inside the cabin can be reconstructed, few passengers if any were looking through the windows, or were aware of the airplane's ominous flight profile—the combination of ultra-low altitude and high speed that characterized its final bombing run. The mood aboard must have been fearful but quieter now. In the back of the airplane another flight attendant was on the phone. Her name was Madeline Amy Sweeney. She had gotten through to an American flight-service manager in Boston, and with exceptional cool had given him a running account of the hijacking, fingering the terrorists and confirming much of Betty Ong's account, including the slaughter of the passenger in business class. It is likely that she added important details about the terrorists' techniques—for instance, how exactly they got into the cockpit or controlled the passengers—which for security reasons have not been made public. Seconds before 8:46 and the impact she looked through a window to give a position report, and to her surprise saw the city flashing by. She said, "I see water and buildings!" She may have been the first person to understand the hijackers' intentions. At the last instant she said, "Oh, my God! Oh, my God!"

Across the East River, in an industrial section of Queens, in a converted Chiclets factory now used as the headquarters of the DDC, Ken Holden had been delayed by the usual round of morning telephone calls. Holden was an observant, quick-eyed man of medium height, with close-cropped curly hair and a body

gone wide after fifteen years of government service. As a registered (if ambivalent) Democrat in Mayor Rudolph Giuliani's Republican administration, he was not a political operative in the
conventional sense but a professional bureaucrat selected for
command because of his willingness to stay in the shadows and
accomplish routine but essential work; he was also known to be
very smart. His intelligence gave him a presence, but not always
to his benefit. He had an acerbic wit that he tried to restrain
when around duller minds. The restraint required effort on his
part. He was by nature amazingly verbose. He had a way of cutting short his sentences by saying "and whatnot," only to go on
immediately to the next thought. Among his subordinates he had
a reputation as a decisive and impatient manager with a strong
instinct for self-preservation. Often he reveled in his job, and
sometimes he did not. He had a badge and an official car with a
driver. He could be arrogant at times, especially with people like
low-ranking cops who got in his way or slowed him down. But
with his inner circle he was unpretentious. He made a comfortable middle-class living, near the top of the city's pay scale, but as
an appointee in an administration nearing its end, he worried
openly about what would come next. He was not a construction
man, and had few ties to the industry. He lived in a modest house
in Queens with his wife, Frances McGuire, an artist, and their
young son, Teddy, named after the first Roosevelt. Holden loved
his family. He liked to read. He drove a Subaru station wagon.
He liked to hike and cross-country ski. He liked his weekends
off.

The agency he ran, the DDC, had been created in 1996 by
the Giuliani administration to oversee the generally thankless
work of building and repairing the municipal infrastructure. Typically the DDC had about 800 projects under way, most of them

very small. With its 1,300 employees and a $3.7 billion design and construction budget, it was a sizable organization, but it lacked the natural political clout of the uniformed services—fire and police—or the prestige of the city's small but highly visible Office of Emergency Management. Giuliani was widely thought to favor all the law-and-order types, because of his background as a prosecutor. This may also simply have been a trait of his personality. He had a reputation for being dictatorial and vindictive. His detractors called him a terror. His admirers said he ran a tight ship. Either way, Holden had learned not to expect friendship from him, and to fear his attention rather than to court it. He had also learned to be on time. The mayor was not expected to attend this morning's 9:15 meeting at City Hall, but his loyalist deputy Tony Coles was standing in for him, and Holden did not want to be late.

Holden tended to sweat when he moved fast. He was heading down the stairs in the atrium of the DDC headquarters when one of his staff members came up to him and reported that an airplane had just hit the World Trade Center. Holden hurried outside to where his official car waited, and in the distance he saw the North Tower burning. One rush then turned into another, and endured.

Were it not for the scheduled meeting downtown, neither Holden nor Mike Burton would have had any reason to get involved in the recovery from the disaster. The DDC was not meant to respond to emergencies. But despite the morning's confusion, the discipline of the schedule prevailed, and without giving much thought to why, each man headed independently toward City Hall—indeed, more intently than before. Actually, Burton was already almost there. He had been coming down the FDR Drive in his Jeep at the time of the first airplane's impact, had noticed the ambulances going by, and had turned on the radio to find out why. After getting stuck in traffic by the Brooklyn Bridge, he pulled to the side, and proceeded through the city on foot. Holden had farther to go, but in a fast sedan with rights to the emergency lanes—a police-package Ford Crown Victoria with a special permit in the front window and a siren and flashing lights. The driver was a short, muscular Dominican named Apollo Hernandez, a bodybuilder who flew small airplanes in his spare time and had a penchant for speed. He was completely dedicated to Holden, whom he referred to as "the Commissioner."

Despite that formality, in the car they had established some of the patterns of a couple. Holden liked to sit in front and give Hernandez driving advice. Today, for once, it did not involve slowing down. As they raced through Queens, they listened to the news radio station WINS, whose tag line was "You give us twenty-two minutes, and we'll give you the world." They lost reception briefly while passing under the East River in the Midtown Tunnel. When the station returned, in Manhattan, the world it gave them was news that a second airplane had come in, and that it had hit the South Tower. The implications were huge. Hernandez launched into an earnest explanation of the fuel capacity of small airplanes. Holden for once was almost silent. Though he had been raised in an observant Jewish family, and as a young man had given faith a serious try for a few months while living in Israel, he had become a thoroughly unreligious person—and also enthusiastically profane. True to himself now, he did not invoke the name of God, as so many others did on both sides that day, but reacted to the magnitude of the event by thinking *Holy shit!* as the car sped downtown.

They parked beside City Hall, about four blocks from the

burning towers. Holden jumped out and said, "Wait here. Don't move. Wait here. I'll come back." Those words haunted him later, after the buildings fell. In the uncertainty of the moment he couldn't help but worry that Hernandez might have followed his orders too faithfully and died. Meanwhile, City Hall was locked tight. Holden gave up on the idea of a meeting, and stood idly outside. Burton had given up on it too, and had boarded a mobile command post—a bus belonging to the Office of Emergency Management—that was creeping west toward the Trade Center fires. Then the South Tower fell. Holden heard it as a growl, Burton as a roar. Both men were caught in the debris cloud, and each moved north among the crowds. When the North Tower fell, Burton was caught in that cloud, too, and Holden was not. Because neither of them was injured, it amounted to a minor difference. Essentially they experienced the same dreamy confusions that were shared by a hundred thousand others in downtown Manhattan.

After a few hours Holden found a functioning phone and first called DDC headquarters, and then called his wife. Later I asked him how she had been. He said, "Upset."

I said, "Crying?"

"No. Very concerned about loss of innocence. 'We've lost our innocence!' she kept saying. I said, 'Try to relax.' I told her I didn't think I'd be home for dinner."

"And when she said, 'We've lost our innocence,' did she mean America?"

"Yeah. Obviously she was watching TV."

He was finally able to contact Hernandez, who, as Holden had feared, had remained loyally parked at City Hall through both collapses, but was unscathed. He had watched the towers fall from close by, had been hit by the successive debris storms, and had lost family in the attack, but he later described the experience to me in the most pragmatic terms, without bewilderment or self-pity. He said, "You go into that alert, that automatic alert. You're just waiting to do the right thing. To me the right thing was to turn on all the flashing lights on the car. In case somebody needed help, they could come to the car. But I wasn't going to move out of there until I heard from the Commissioner."

Hernandez was thirty-nine years old. He had immigrated to New York from the Dominican Republic at age thirteen, in December of 1975, and despite the cold weather had immediately felt at home. He had become a citizen in 1994. He had led exercise classes at a Jack LaLanne gym, and had met his wife there. He had taken a job with the city for the security it offered. He was a good worker. He was willing and uncomplaining. He was very clean-cut. He did not drink or smoke. He had three little children, and was a good father. Now he was the perfect soldier, an American ideal. While hundreds of panicked people went running by, some in the wrong direction, he stayed at his post by his Ford Crown Victoria, and without fear.

Two women rushed up to him in terror, hugging each other and sobbing. "Oh, my God, the towers fell!" Hernandez calmed them down. They walked away. He spotted a couple of wounded policemen on their hands and knees, one of them bleeding from the head. He got them into the back of the car, and gave them a blanket with which to wipe their faces. They called him their guardian angel. The dust cleared. Attracted by the flashing lights, a woman approached who was shaking all over and could not speak. Hernandez found help for her and returned to the car. He noticed that his cell phone and radio links had been disrupted. After ninety minutes in position, he realized that Holden would no longer be looking for him here. He drove the policemen to the vicinity of a first-aid station, and then headed slowly uptown. Eventually the cell phone began to function, and Holden got through to him. They had a short, happy reunion on Carmine Street, in Greenwich Village. By early afternoon Hernandez had parked downtown again, and like a good soldier was waiting close to where he had waited before. Holden and Burton had met up in the police headquarters nearby, where a temporary command center had been established, and they were trying to marshal the DDC's resources to help. Hernandez wanted to help too. He was not tired or bored. He was patient and proud. He felt love for his adopted country. He had never been in combat, but understood waiting as the experience of war. The streets were mostly quiet. Heavy smoke was rising into the sky.

H olden and Burton responded tactically, with no grand strategy in mind. At the police headquarters they discovered a telephone in a room off the temporary command center—a chaotic hall filled with officials struggling to get organized—and they began making calls. No one asked them to do this, or told them to stop. One of the deputy mayors there had formally been given the task of coordinating the construction response, but with little idea of how to proceed, he had so far done nothing at all. Holden and Burton themselves were operating blind, groping forward through the afternoon with only the vaguest concept of the realities on the ground. The DDC's previous experience with emergencies had been limited to a sinking EMS station in Brooklyn, caused by a water-main break, and a structural failure at Yankee Stadium, one week before baseball's opening day. Without yet having visited the site, Burton could not now bring himself to believe that the Trade Center towers had completely collapsed.

Imagining the need to protect people from falling glass, he phoned a scaffolding company and requested that they prepare to load a half mile of sidewalk bridging onto their trucks. Holden later told me this seemed to him like taking a Band-Aid to a gunshot wound—but at the time there were more important things to do than quibble. It seemed likely that hundreds of people lay trapped in the ruins, suffering and slowly dying. Holden arranged for a police escort to bring in twenty-five light towers from Queens for the coming night. Burton established an expanding "phone tree," alerting the city's heavy-construction industry to ready itself for emergency requests later in the day.

As might be expected, the industry was something of an intimate circle, in which many of the top people had known each other for years. Burton himself was one of them, though he was currently engaged in a foray through government service. Faced with the urgent need to get crews and heavy equipment onto the job, he bypassed ordinary bidding procedures and made some immediate choices on the basis of personal and corporate reputations, asking AMEC, Bovis, and Turner to send representatives to an initial meeting that afternoon at police headquarters. Holden brought in Tully as well. As it turned out, these were the four companies that ultimately would share the lucrative job of cleaning up, earning profits in the millions, though at considerable exposure to potential claims and lawsuits, because neither the companies nor the city were ever able to obtain adequate insurance for the Trade Center project. Cynics who later implied that the choice of these companies was in some way an insiders' deal were only superficially right. The four companies were simply the first that came to mind—and on the day of the collapse they responded altruistically in the face of enormous confusion, thinking at the most a few days ahead, without even the possibility

of calculating gain. Holden and Burton also brought in the
renowned engineering firm of Thornton-Tomasetti. Soon afterward the firm's habitually well-dressed president, Richard
Tomasetti, met up with some of the other chosen representatives, and rode in an escorted van to police headquarters downtown.

The first meeting didn't amount to much, in part because police officials refused to give the group access to the site, forcing
the construction men to begin planning their response on the
strength of television images and secondhand news. It was understandable if the police were perhaps overly rigid, or were operating on the basis of suddenly obsolete concepts of public
safety: they had seen thousands of citizens killed just a few hours
before, and had lost twenty-three of their own people, and for
the very reasons they had chosen the profession, many of them
must have had great difficulty with America's apparent loss of
control. Nonetheless, Holden and Burton urgently believed that
their group had important capabilities to offer the search-and-
rescue effort, and they felt increasingly frustrated as the day
dragged toward night and they continued to be blocked. They
also felt emotions that Holden could not admit to then but later
described to me as only human—an urge to get to the center of
the action, and a powerful curiosity about conditions there. The
two men kept demanding to go in, and despite a short setback
when World Trade Center Seven collapsed, they finally wore the
police down. At 5:30 that afternoon they got permission for their
nascent team of unbuilders to explore the ruins firsthand.

The walk they took became famous at the site because of the
forces it unleashed. About fifteen people went along, including
Holden and Burton, Richard Tomasetti, and a collection of tough
construction guys. Though every one of these men was accus-

tomed to grappling with problems on a very large scale, none was prepared for a disaster so vast and severe. Wading through paper litter and pulverized concrete, they tried to approach from the north, but were blocked by heavy smoke. Few of them had respirators. They moved upwind to the Hudson and flanked southward again, cutting through the peripheral public areas of the heavily damaged World Financial Center, a high-rise complex that stood between the Trade Center and the river. Conditions there were so strange that Holden afterward had trouble knowing where they had been, though he did remember one surreal passage through a ruined atrium restaurant where all the fire alarms were blaring, *wah, wah, wah.*

Then, through an opening between buildings, they came suddenly upon the Trade Center ruins themselves—the skeletal walls and smoking hills of rubble where the towers had been, the boxy shell of the Marriott hotel, the heavy steel spears protruding from neighboring buildings, the collapsed north pedestrian bridge, the massive external column sections thrown every which way across the streets, and everywhere the fires. For Holden it looked like a scene from *Apocalypse Now.* He told me, "It was hallucinogenic—quasi-druggy, with flares shooting up and death in the air. There was a sense of crazed panic, people fighting to save lives, fire hoses cascading all over the place." For the next ninety minutes they moved through a smoky twilight among the ruins. The ground was littered with hundreds of shoes, presumably from victims, but characteristically for this unusually imploded killing zone, not a single corpse lay in sight. Afterward Holden was nearly at a loss for words to describe what went on in his mind. He said, "Cars. I remember everywhere you'd walk there were crushed cars. Some that had burned. Sparks flying. Flames shooting. Smoke all over the place. People desperately trying to move the debris. The light was dim. It was just this really odd, frantic atmosphere. We were walking through twisted debris. It was completely startling—like a 'Holy shit!' experience. Every corner you turned was 'Holy shit!' "

Mike Burton and Richard Tomasetti stayed together, each struggling to maintain an analytical frame of mind. It helped to confer with each other. Their focus was on lifting the steel to uncover survivors. It was obvious that large numbers of skilled workers were necessary, along with more heavy equipment than the two men could have imagined before. They needed not just the largest diesel excavators and cranes to pick up the pieces but, for lack of stockpiling space, a fleet of trucks to haul off the debris—and they needed it all right away. The first problem was access: a major effort would be required to clear a path down West Street just to get equipment to the ruins of the north pedestrian bridge. That bridge by chance was a Tomasetti design, and it was stout, but it had been hit with impossible force when the North Tower fell. Now it was blocking the best route to the pile.

The group moved south under the crackling flames of a burning building, a 1907 landmark at 90 West Street, and they skirted a severed 767 wheel before heading back toward the center of the site by the severely wounded Bankers Trust building. Suddenly a fireman rushed up to Burton and Tomasetti and said, "Listen, there's a water-main break, and there's no pressure to fight the fires. What we want to do is tap into the water tower at the top of this building." He pointed to Bankers Trust. "I want to know if it's safe."

Was it because of their office clothes? The fireman did not know Burton or Tomasetti, but had read them correctly as engineers. Burton looked at Tomasetti, who peered at the building through the smoke, and hesitated. The structure had been speared and torn open. The fireman mentioned that his people had already been in the basement, searching for victims. Tomasetti said, "Well, it's kind of dark out, but the building seems to be missing only one column . . ."

As the second-in-command of an obscure city agency in Queens, Burton had not the slightest authority here at the Trade Center site, but he was willing to assume it anyway. This was typical of him. Burton's brashness was well known within the DDC, where people tended to be cautious of him for his disdain of procedure, and for what many saw as his arrogance. Holden himself was wary of Burton, and said he felt the need to "cage" him regularly, but he recognized Burton's effectiveness nonetheless. Burton was the DDC's doer. And Burton had an agile mind. He was a right-wing conservative but also surprisingly supple, and he understood now that conventional standards of safety simply did not apply to these ruins. To the fireman he said, "Listen. You've already been in the building. You've already risked your life once. It would take us a week in daylight to tell whether the building is safe or not. If you climb the stairs on the far side, it's probably okay. That's a guess. But it won't be any riskier than where you've already been." With that the fireman murmured into a walkie-talkie and disappeared. Burton exchanged glances with Tomasetti. He saw a group of firemen go running toward the building, and wondered about the wisdom of his advice. He did not waffle publicly, however, and it later turned out that he was right. Bankers Trust did not fall—and indeed, not a single person was killed during the nine months of the recovery effort that ensued. Burton had just made the first of his many quick decisions at the Trade Center site. He knew already what others soon discovered—that he had a particular talent for making up his mind.

The group returned to police headquarters after dark and

worked through the night, mobilizing columns of equipment and thousands of skilled laborers, considering the rudiments of a disposal program, laying out the role of engineers, planning the first rapid inspections of hundreds of neighboring buildings, and outlining the four-quadrant overlay that would divide the excavation work according to the simplest geometry. The four-quadrant pattern owed as much to the presence of the four companies on that evening's tour as to a compelling operational logic, but while it endured, during the first urgent months, it functioned reasonably well. There was no time for refined planning anyway. Holden went home, and got two hours of sleep; Burton spent fifteen minutes dozing in his Jeep. Then they returned to the site and continued to work tirelessly. They were bit players at first, and did not pretend otherwise: the mayor's Office of Emergency Management was supposed to be coordinating things, and the main response was overwhelmingly and appropriately the Fire Department's. But to a degree that was surprising even to Holden and Burton, over the next few days they were able to bring New York's enormous construction energies to bear. As individuals, they remained almost invisible, but their influence rapidly grew.

None of this reflected the normal operation of the U.S. emergency-response system. After natural and man-made disasters in the past—notably the 1995 bombing of the Murrah Building, in Oklahoma City—the efforts on the ground had rapidly been nationalized under the direction of FEMA and its operational allies in the Army Corps of Engineers, a branch of the military that calls itself the world's largest public engineering, design, and construction-management agency. These people in turn dealt primarily with pre-assigned companies that specialized in the disaster-cleanup business (many based in Florida, along the hurricane tracks). The system was intelligently organized and charted—and given the frequency of destructive weather in North America, it was well exercised. It was not, however, quite prepared to operate with a speed and intensity equal to New York's.

Ken Holden, for one, didn't even consider turning to the outside for help. He was vaguely aware of FEMA and the Army Corps of Engineers, but with no previous experience in national emergencies, he neither knew nor wondered about the protocol. He might have assumed that ultimately the federal government would pay the bill, but afterward he had no memory that such questions crossed his mind. To me he said, "None of us wondered, 'Should we contact the state? Should we contact the feds? FEMA? The Army Corps?' It was just 'We've got a disaster here. Let's fix it.' It was instinctive. 'Let's give the Fire Department access. Let's let the police see if they can rescue their own. This is not out of our realm.' "

By "realm" he meant professional expertise, but he might as well have been talking about his beloved city. He was a typically proud New Yorker. He said, "As I kept explaining to FEMA later, this is not Oklahoma. We had the equipment. We had the connections. We could handle it. We just went in and did what we had to do. And no one said no."

"Did anyone say yes?"

"No. But then again, we weren't asking."

More than pride was at play: Holden simply did not have the time to seek permissions. The rush to find survivors was hopeful at first, and then less so. Sam Melisi later described for me the wild-eyed urgency of the initial search. Speaking of the lost firemen, who throughout the months to come provided a focus to the recovery efforts, he said, "These were people you had worked with, and they were maybe alive. You knew they were trapped in there, and there was a sense of franticness, and it was personal. I remember crawling through the steel—it would have probably been by the hotel. There were some spaces that let you get below and take a look around. It wasn't regulated at all. The first couple of days, anything went. It wasn't like somebody was saying, 'You can't go in there, you can't do this, you can't do that.' It was more like 'Hey, if you think you can get in there, go ahead.' All bets were off. It was just 'Go and bring somebody home.' "

At age forty-three, Melisi was a small, wiry man who had a disarming way of suggesting his opinions rather than asserting them. He had a nasal voice and a big moustache. He was obviously somewhat shy. Within the wolf-pack world of the Trade Center site he became known at first simply because his diffidence was unique. Still, he was a fireman through and through, with strong allegiances to the department and a blue-collar history that was fairly standard for the type. He grew up on Staten Island as the son of a diver with a small marine-salvage business, and he excelled in high school before heading to college in Oregon, where, cut off from his roots, he floundered. After dropping out of school, he hired on at a sawmill for six months, and then returned to New York, went to work in heavy construction, and eventually became a licensed equipment and crane operator. At the advanced age of twenty-seven he joined the Fire Department because of the generous time off ("Like being on permanent vacation," he told me) and only then discovered the job's power to mold people's lives, including his own. So far he had put sixteen years into the service—in various ladder and rescue companies, and most recently as an assistant in the engine room of the city's 1938 fireboat, called the *Firefighter*. Because of his heavy-equipment experience and additional training, he also served on a specialized collapsed-building team, which had responded to the 1993 bombing of the Trade Center and had been dispatched to the subsequent bombing in Oklahoma City and to a hurricane disaster in the Dominican Republic. He lived with his wife and two young children on Staten Island, in a surprisingly rural setting—a small wooden house from the 1840s with a large back yard bordering on a forest preserve and littered with old construction equipment, including a small crane. (One day at the site he said, "People here keep saying how strange it all looks, but I dunno, it kind of looks like my back yard.") In a shed at the end of his yard he had built a welding and machine shop. The jobs he did there helped him to supplement his modest salary. Sometimes he moonlighted as an electrician. Sometimes he was a plumber, sometimes a carpenter, too.

When the first airplane hit, Melisi put down his reading in the *Firefighter's* engine room and prepared to get under way. When the second airplane hit, he understood it meant war, and he had the strange impression of feeling every possible emotion all at once. When the South Tower fell, the boat was plowing at full speed across New York Harbor, and the twin diesels were roaring. When the North Tower fell, the boat was pulling up to shore, and the diesels were still so loud that Melisi did not hear the thunder. The boat docked just north of the North Cove, along a riverfront promenade. Melisi emerged from the engine room and helped the deck crew drag hose to the fire-suppression teams on the northwest corner of the site. He returned to the boat and helped to engage its pumps, which delivered enormous quantities of harbor water to the ruins over the following two weeks. Aware of Melisi's training in collapsed buildings and rescue operations, his supervisor then cut him loose. He said, "Just go. See what you can do." This turned out to be more than a little thing. Melisi's maritime career had come suddenly to a halt.

He joined the scramble through the smoke and debris, searching for cavities in which people might have survived. This was a collapse unlike any he had seen before, and not only in size. Though the top layer of the pile was jagged, it was also fantastically dense, and it offered little in the way of natural shelters, or of access to the underground. Melisi circled to the north side, where fires raged and Building Seven threatened to fall, but the devastation was less severe. Handicapped by the lack of a flashlight, he joined a rescue team that descended through a crater in Building Six and made multiple attempts on the North Tower basements, but was blocked each time by smoke and heavy debris. Back on the surface again, he moved down West Street, squeezed under the ruins of the north pedestrian bridge between the hulks of crushed fire trucks, and spent the afternoon with a shovel, digging for survivors and uncovering only the dead. He arrived in the site's southwest corner that evening at about the same time that Ken Holden, Mike Burton, and the unbuilding crew first came walking through. Melisi neither knew nor noticed them then, but he shared some of their construction expertise, and he had drawn conclusions similar to theirs—that the bucket brigades he saw operating on the pile were ineffectual, and that if victims were to be found alive, it was essential to bring in large cranes and grapplers that could lift the fallen steel columns and expose cavities below the surface. Typically, Melisi did something practical about it right away. He worked to clear an access route from the south, directing front-end loaders to shove cars and toppled lampposts aside, opening a path that detoured around the heaviest obstacles. A perverse pattern then prevailed: the fast-moving small equipment was the first to arrive, and time and again it had to be dismantled or laboriously moved out of the way in order to allow the larger and more effective pieces in; because of the quantities of debris, space was simply not available for both. The delays were frustrating for Melisi, who through his actions and expertise was unintentionally already involved in the management of the site. The greater frustration came later, however, and it was the near total lack of survivors.

The lack was partly definitional, since it excluded the 15,000 people who were able to walk away, in some cases even after the buildings fell. Their survival, however, did not diminish the reality that thousands remained unaccounted for, or the terrible feeling in those first confused hours and days that time was running out. It went without saying that the survivors who mattered were the ones who might now be lying trapped in the debris. They turned out to be rare: over the subsequent months of retrieval it became obvious from the condition of the bodies that few if any of the victims had perished while awaiting rescue. By the final count, in a place where nearly 3,000 had died, only eighteen people were recovered alive. Two of them were injured policemen discovered on the first day in the underground concourse, a shopping area east of the South Tower that had been speared and pummeled by falling columns but not completely crushed. The remaining sixteen were all found among the ruins of the North Tower. Fourteen of them—twelve firemen, one policeman, and one civilian office worker—came through largely unscathed in an intact stairwell section between the second and fourth floors, sandwiched between collapses. When they were rescued, also on the first day, they emerged from the ground as if from the bowels of hell, and cheering broke out across the site.

But only two other people were ever found alive. Both were

Port Authority employees caught at relatively high elevations in the North Tower. They did not "surf" the collapse, as a couple of Port Authority cops were falsely rumored to have done in the South Tower, but they lived through falls that should have killed them. The first of them was a thirty-two-year-old staff engineer named Pasquale Buzzelli whose job involved overseeing work on the George Washington Bridge—including Rinaldi's project of decorating the span with lights. Buzzelli was riding an intermediate-stage elevator to his office, on the sixty-fourth floor, when the 767 slammed into the North Tower far above. The elevator shook violently and briefly dropped before catching itself and returning slowly to its starting point, the "sky lobby" on the forty-fourth floor. When the doors opened, Buzzelli was confronted by a confusion of shouts and thick black smoke (presumably from burning jet fuel that had poured down other shafts). He retreated into the elevator and, with no way to go but up, instinctively pushed the button for his familiar sixty-fourth floor. It was a slow ride. When he got there, the floor looked well lit and calm, and it was almost smoke-free. Most of the workers had already left, but more than a dozen remained, and were dutifully awaiting instructions from the authorities below, as apparently they had been asked to do by Port Authority and Fire Department officials on the phone. Because the group included Buzzelli's supervisor, a man named Pat Hoey, Buzzelli decided to stay too. He said, "What happened, Pat?"

Hoey said, "I don't know, but I was just about thrown out of my chair."

"Really? I thought it was some kind of elevator thing."

Hoey kept calling downstairs to the Port Authority communications center and was unable to get clear information. He seemed edgy, but more from frustration than fear.

Buzzelli got on another phone and called his wife, who was seven months pregnant with their first child. He said, "Don't be alarmed or anything like that. I'm okay. Just put on the TV. Tell me if something has happened."

He waited. After a while she came back and said, "Oh, my God, Pasquale! There's a plane in your building!"

Buzzelli lived up to the reputation of engineers as unflappable. He said, "All right, all right. Don't get excited." (When he recounted the conversation later, he explained, "She was getting all worried and stuff.") He said, "Just can you please tell me where it is? Did it hit high, middle, or low?" ("Because she doesn't really know the floors and stuff.")

She said, "Well, it looks pretty high up in the building."

This was of course good news. Buzzelli had been through the 1993 World Trade Center bombing, and was not overly concerned. He said, "Okay. Well, just so you know, I'm okay, and we're here, and we're going to figure out what to do."

It took them a while. They learned that the South Tower, too, had been hit by an airplane—but they neither heard nor saw the impact. Their offices occupied the northwest corner of the North Tower, as far from the South Tower as could be. They were surprisingly isolated, and were unaware that their upstairs neighbors were jumping from windows and falling by outside. They stuffed wet coats under the doors against the faint smoke that was drifting through the elevator lobby. Apparently their reactions were slowed by a sort of collective inertia: believing they had been told to stay, they worked the phones looking for someone who could give them a schedule for the evacuation. They delayed for more than an hour, during which Frank Lombardi, who would have told them immediately to go, climbed down the stairs from his office overhead, but of course without checking in on every floor.

At 9:59 they heard muted thunder and felt the building vibrate. It was the South Tower collapsing, an unseen and unimaginable event. They attributed the commotion to something less, maybe a piece of airplane breaking free and sliding down the outside. They did not go to that side of the building and look through the windows. However, they did notice that the smoke in the elevator lobby was growing thicker. Buzzelli and another man unblocked the doors and went out to check the nearest stairwell, which they found clear and well lit. They returned to the office and reported the news. The time had come to leave, with or without permission. At last the group started walking down from the sixty-fourth floor.

There were sixteen of them. They moved at different speeds and eventually spread out over at least nine floors. The stairway descended with left turns, in a counterclockwise direction, and it was of course windowless and completely cut off from the outside. About a third of the way down (in the forties) Buzzelli began to encounter exhausted firemen, some of whom were sitting on the steps and resting. They knew no better than he that the South Tower had fallen or that their forces had been ordered to retreat. They were calm, and said, "Just keep going down, clear run. Keep going down, clear run."

Buzzelli had just passed the twenty-second floor when the North Tower gave way. It was 10:28 in the morning, an hour and forty-two minutes after the attack. Buzzelli felt the building rumble, and immediately afterward heard a tremendous pounding coming at him from above, as one after another the upper floors collapsed in sequence. Buzzelli's memory of it afterward was distinct. The pounding was rhythmic, and it intensified fast, as if a monstrous boulder were bounding down the stairwell toward his head. He reacted viscerally by diving halfway down a flight of stairs and curling into the corner of a landing. He knew that the building was failing. Buzzelli was Catholic. He closed his eyes and prayed for his wife and unborn child. He prayed for a quick death. Because his eyes were closed, he felt rather than saw the walls crack open around him. For an instant the walls folded onto his head and arms and he felt pressure, but then the structure disintegrated beneath him, and he thought, "I'm going," and began to fall. He kept his eyes closed. He felt the weightlessness of acceleration. The sensation reminded him of thrill rides he had enjoyed at Great Adventure, in New Jersey. He did not enjoy it now, but did not actively dislike it either. He did not actively do anything at all. He felt the wind on his face, and a sandblasting effect against his skin as he tumbled through the clouds of debris. He saw four flashes from small blows to the head, and then another really bright flash when he landed. Right after that he opened his eyes, and it was three hours later.

He sat up. He saw blue sky and a world of shattered steel and concrete. He had landed on a slab like a sacrificial altar, perched high among mountains of ruin. He was cut off by a drop of fifteen feet to the debris below him. He saw heavy smoke in the air. Above his head rose a lovely skeletal wall, a lacy gothic thing that looked as if it would topple at any moment. He remembered his fall exactly. He assumed that he was dead. He waited for a while to see if death would be as it is shown in the movies—if an angel would come by, or if he would float up and see himself from the outside. But then he started to cough and to feel pain in his leg, and he realized that he was alive. He was trapped high on the altar, injured, and covered with a slick powdery dust. He shouted for help and called out the names of the people who had accompanied him down the stairs, but heard only silence in response, and saw no movement of a human kind. Where the Twin Towers should have stood he saw only smoke and sky. Somehow an entire huge building had passed him on its way to the ground. Somehow also he had landed just right. Buzzelli was a Catholic, but an engineer, not a theologian. For an hour he sat trapped on the altar trying to reason things through.

Conditions were still precarious. The altar itself was unstable, threatening to capsize or collapse. Buzzelli was too badly injured to climb down. He worried that if he rolled off his perch—simply allowed himself to drop—he would be impaled on the twisted steel below. Increasingly he had trouble staying calm. He was alone. He was helpless. And when the silence was finally broken, it was not by the sounds of rescue, but the crackle of an approaching fire. The fire came from an area behind him, which from his position he could not see. He worried that the fire would weaken the skeletal wall and cause it to topple onto him— or, worse, that the flames would burn him alive. Judging from the sound, the danger was growing rapidly.

Then he heard someone call, "Richie!" It was a fireman climbing unseen through the rubble nearby, attempting to locate the fourteen stairwell survivors, one of whom had established radio contact and was guiding in the rescuers.

Buzzelli did not know this. All he knew was that there was a human presence. He started shouting, "Help, help! I'm up here!"

Eventually the fireman materialized below the altar. He looked up at Buzzelli and said, "Oh. Do you need a rope to get down?" He seemed to think Buzzelli was a fellow rescuer who somehow had gotten stranded on top of the slab.

Buzzelli said, "I'll jump if you want me to, but I can't climb down alone."

Apparently, this was not what the fireman expected to hear. He did a double take and said, "Oh, my God. Guys, we've got a civilian up here!"

Other firemen arrived and began to discuss how to get Buzzelli down. It was not an obvious thing. They were worried about the instability of the debris slopes as well as the precariousness of Buzzelli's perch. They were also worried about the fire. It was not just a fire, it was an inferno. Indeed, just then the flames flared up, and came on so aggressively that the firemen had to retreat. As they disappeared, one of them yelled, "Hold on! We'll get back to you!" The promise was of little comfort to Buzzelli, who figured, probably correctly, that they would not have left him if there was any way they could have stayed.

Buzzelli by now was thoroughly terrified. The fire was roaring, popping, and setting off small explosions just behind the altar. He still could not see the flames, but he could feel the heat. Now he heard a new, more intimate sound, which he took to be the groaning and sizzling of overheating steel. This was the end. Desperate at least to assume control over his fate, he groped around and found a sharp metal scrap with which to slash his wrists. He had it firmly in hand and was about to cut himself open when, strangely, just as suddenly as the fire had grown, it subsided and died. A few minutes later the firemen reappeared. One named Jimmy said, "All right, we'll get to you somehow." He circled clockwise around the altar, disappeared for a stretch of dangerous climbing, and pioneered a route to an alcove high in the rubble mountains above and behind Buzzelli. Somehow he clambered down to the altar. Three others followed. The firemen then fashioned a rope cradle, got it around Buzzelli, and lowered him to the debris slope below.

The group still had 400 yards of difficult terrain to go. Despite the severe pain in his leg, Buzzelli managed to walk about halfway before beginning to lose consciousness. The firemen put him onto a plastic stretcher known as a Stokes basket, and they passed him down the pile in the manner of the bucket brigades.

In the ambulance a kindly attendant lent him a cell phone to call his wife. She was at home, surrounded by friends and family. It was late in the afternoon. She said, "Oh, my God, I can't believe . . ." He heard an uproar in the background. He said, "Yeah, I'm alive." He was taken to Saint Vincent's Hospital, where the other patients were firemen or cops—rescuers who had been lightly injured in the debris. The staff assumed that Buzzelli was just another one of them. He had some cuts and bruises, and a broken right foot—that was all. They told him to go home or sleep in the cafeteria if he liked, because they were still thinking triage then, and standing by for the rush that never occurred.

Also with Buzzelli in that stairwell was Genelle Guzman, a thirty-one-year-old Port Authority clerk of Trinidadian origin, who had delayed with the others on the sixty-fourth floor, but had gotten ahead in the descent, and was on the thirteenth floor when the building boomed and broke apart around her. She was with a friend named Rosa, and had stopped to adjust her shoe. She had put her hand on Rosa's shoulder. As the building crumbled, she felt the shoulder pull away, as if Rosa were running up the staircase. She was hit, and felt the acceleration of the collapse around her. All was darkness. When the roar stopped, she heard a couple of calls from a man who then grew silent. Her head was jammed under a load of debris, but eventually she worked it loose. She was in a dark cavity in the inner world of rubble. Her legs were pinned and crushed. She felt a dead man beside her. It turned out that he was a fireman, and that there was another one, also dead, lying nearby. For twenty-seven hours Guzman lay trapped and seriously injured. She spent some of that time bargaining for her life, pleading with God to show her a miracle. Early the next afternoon she heard a search party, and when she yelled out, a voice answered. The voice said, "Do you see the light?" She did not. She took a piece of concrete and banged it against a broken stairway overhead—presumably the same structure that had saved her life. The searchers homed in on the noise. Guzman wedged her hand through a crack in the wall, and felt someone take it. A voice said, "I've got you," and Guzman said, "Thank God." She spent the next five weeks at New York's Bellevue Hospital, undergoing reconstructive surgery on her right leg. She was the last person to emerge alive from the ruins.

Of course, that was not at all obvious on September 12—and probably even less so at the site than elsewhere. Experience with earthquakes and earlier bombings suggested that a significant percentage of people caught in the collapses would still be alive. This had been true even in the ruins of Hiroshima and Nagasaki. Sam Melisi said, "All we knew was that time was of the essence. It was just really important that we get to people as soon as possible, because the longer we delayed, the less the chance of survival was. We knew that from other collapses we'd gone into."

I asked him an academic, almost nonsensical question. "What's the rule on that? Is there a known interval?"

Melisi gave me a typically practical answer. He said, "Sometimes people write about the 'Golden Hour,' and this and that. But it depends on their injuries." Are they crushed? Can they breathe? By any chance do they have water? Victims have been known to last a week trapped in rubble. Victims have been known to last longer. Melisi said, "It was hard to be sure of anything. We knew we were looking for people, and we knew they were there somewhere."

It was a grim sort of hope, therefore, rather than specific

knowledge, that inspired the early rush at the site. Having lost
its headquarters bunker with the collapse of Building Seven,
the city's Office of Emergency Management was showing surprising flexibility—a mature recognition by some of its senior staff
members that their plans and procedures had been overtaken by
events, and that their organization lacked the capability to direct
the response at the site. They knew that their most useful role
now would be to help with some of the peripheral logistics—
office space, food sources, press briefings, and so forth—and to
coordinate between the myriad governmental agencies that were
starting to crowd forward and jump into the fray. With those fi-
nite goals in mind, the OEM moved fast. By the second day it
had set up an emergency command center in the ground-floor
cafeteria of PS 89.

For the first several days the scene in the cafeteria was fren-
zied. A marine-construction manager named Marty Corcoran
told me early on about his attempts to make sense of the chaos.
Corcoran was a tall, powerfully built man who looked as if he
might enjoy crumbling concrete with his bare hands, but who
turned out to be a sensitive observer of the Trade Center scene.
He had arrived the day after the attack on behalf of his employer,
Weeks Marine, a New Jersey–based concern that, having rallied
in heartfelt reaction to the disaster, had also quite rationally understood the need to set up a big barging operation. In the long
run the company's motivations were probably about typical for
the Trade Center site—neither as altruistic as they had been
over the first few days, nor as mercenary as cynics later implied.
After three hours of wandering the site in astonishment and confusion, Corcoran found his way into the headquarters at PS 89,
where he assumed he would discover a traditional chain of command. He said, "It was unbelievable. There are ten thousand

meetings going on. Equipment is being mobilized to different places. The OEM has lost everything—it's like all their preparations were thrown out the window. Now you have OEM, you have the Fire Department, you have EMS, you have Salvation Army, you have all these billions of people involved, and you're sitting at a cafeteria table trying to discuss something with all these distractions going on around your head. It was insane."

Corcoran heard that the site had been divided into quadrants, and that they had been assigned to four companies—every one of which struck him as an unexpected choice. To me he said, "Who's AMEC? Who's Bovis? Oh, we knew Turner, but they're construction managers—they don't even do the work themselves. The only company here with equipment was Tully, and Tully is well connected with the city, I guess. That's how they . . ." He hesitated, worried possibly that I would interpret this as an allegation of impropriety, which in context it certainly was not. He said, "I never quite understood how it happened, but it was kind of moot at that point. If these were the rules of the game, you just had to figure out how you were going to play by them." It turned out that the only rules that mattered were those related to speed and physical progress, and Corcoran was good at them. He heard about an obscure city agency called the DDC that for unknown reasons was emerging in an unscripted leading role. A couple of guys there were said to be making decisions. Apparently the one to see was a man with a moustache and narrow shoulders, named Mike Burton.

Corcoran found Burton under full assault on the far side of the cafeteria. Later he said, "He was just being bombarded by people. I tried to explain to him what we do and don't do. I can remember the look on his face. It was like . . ." Corcoran pondered the memory. "Like he was a deer caught in headlights.

Like 'Where's the next shot coming from?' And I felt embarrassed, because it was like I was speaking another language to him. The last thing he wanted to hear was some marine contractor coming in with ideas and schemes."

But the scheme Corcoran had in mind was important. Already by that second day huge volumes of steel and debris were emerging from the site as AMEC fought its way down West Street toward the north pedestrian bridge, and Bovis came in from the south, both companies working furiously to clear room for the cranes and heavy equipment that might help in the search for survivors. Tully was cranking up too, over in the southeast corner. Something had to be done with the resulting material, and right away, or it would start to clog the rescue operation. But this was not Oklahoma City, where cattle still graze close to downtown, and empty space is everywhere. In New York there was neither the space to pile the stuff nearby nor the time to go looking for suburban pastures. Confronted with the need for an immediate solution, the Giuliani administration decided to reopen the city's recently retired Staten Island landfill—a famously unpopular dump more than twice the size of Central Park, which rises along a saltwater channel and is known (from the Dutch) as Fresh Kills. City trucks began rolling there the very first night, each carrying its dusty little load along a congested route through much of New York—by tunnel to Brooklyn and expressways across the borough, then over the Verrazano Narrows Bridge and down the length of Staten Island. One way alone, that trip could take two hours or more. What Corcoran proposed instead was using barges—equipment that Fresh Kills was already set up to accommodate. The use of barges would keep the Trade Center ruins off the roads, and limit the spread of noxious dust. More important, each barge was capable of carrying fifty to a hundred truckloads of debris, and the barges could be lashed into rafts of four before being pushed across the harbor and up the channels, twenty-six miles to Fresh Kills. The capacity of the barges was so large that with merely two loading points in Lower Manhattan, Corcoran believed he could keep the transportation of the debris from ever bottlenecking the efforts at the site. The loading points would have to be dredged to give the barges sufficient depth at low tide—and because of environmental regulations, this would require permits. But Corcoran had tugs and dredges standing by, and he could set up the operation fast.

For a deer caught in headlights, Burton was impressively decisive. It turned out that he had heard and considered everything Corcoran had said. He asked him to start setting up the barge operation right away, and to get back to him as soon as possible on the permit process. Nonetheless, Corcoran's memory of Burton's condition over those first few days was about right. Burton was functioning almost entirely without sleep, and reeling at the increasingly evident size and complexity of the project that he and Ken Holden seemed somehow to have taken on. With several thousand people missing inside the rubble, and volunteers still swarming across the burning pile, the pressure was relentless—and there was no end of it in sight. For reasons of personality, Holden and Burton had not had an easy relationship over the previous few years, but in the cafeteria now they forgot their differences and stood back-to-back, inventing solutions to problems as they arose, and sharing the simultaneous feelings of urgency and fatigue that characterize the battlefield experience. They were not the only ones. After a few days at PS 89, Corcoran, too, was walking around with a thousand-yard stare.

There was a significant change after three days when Holden and Burton left the cafeteria crowds, and moved their operation

one floor up to the kindergarten space—the incongruously childish classrooms, with their alphabet posters and diminutive chairs,
from which the two men would direct the operation for months
to come. The mood there remained as urgent as before, but because the clamor was reduced, a raw form of organization was
able to take hold. It happened almost spontaneously. Holden and
Burton assembled a small team from DDC headquarters in
Queens, and supplemented it with a few outsiders—notably Peter Rinaldi, the Port Authority engineer, and a hard-nosed construction executive named Bill Cote, who had been Burton's
roommate in college, and arrived now to serve as his right-hand
man.

No one had time to ponder options and write plans. It was action, pure action, that was called for. Because of the need for
clear communication, Burton instituted large twice-daily meetings in one of the kindergarten rooms—a simple, low-tech management system that proved to be particularly well suited to the
apocalypse outside. Burton's reasoning was lucid as usual. To me
he said, "The only way we can get control of the situation is by
having everyone here. There's no time for distributing memos or
waiting for the chain of command. Everybody has to hear what
the problems are. The decisions have to be made, and everybody
has to hear those decisions. We have to keep everybody moving
in the same direction."

The men and women who crowded in every morning and
every night were the envoys of their respective organizations.
They came to the meetings in groups of two or three. At least
twenty government agencies were represented (in an alphabet
soup of acronyms), but primarily the participants were construction men, a few wearing suits and ties, but most in the industry
standard of jeans or brown canvas coveralls, and steel-toed lace-

up boots. They had helmets and hardhats and yellow Rite in the Rain all-weather field books, and by the second week they were carrying astonishing quantities of cell phones, beepers, and two-way radios—some of which, inevitably, had not been muted.

"Hey, Vinnie, where ya at?"

The participants sat on the windowsills and kindergarten desks, and crowded around the cluster of folding tables at the center of the room. The meetings were informal, and unusually frank. To encourage honest expression, electronic recording devices were banned, and the only minutes kept were of a sketchy, checklist variety—essentially just a schedule of problems to be solved. People were expected to propose solutions even at the risk of seeming foolish, and to swallow their pride when their ideas were dismissed or their performance was bluntly criticized. This happened often. Some of the participants were accomplished people with impressive résumés, but within the inner world of the Trade Center site it hardly mattered what they had done before. However temporarily, there was a new social contract here, which everyone seemed to understand. All that counted about anyone was what that person could provide now.

Mike Burton usually arrived at the meetings last of all, accompanied by the DDC team from the kindergarten room across the hall. He presided over the room by perching on the seatback of a folding chair, giving him a slight advantage of height over people who were seated at the tables. He kept his yellow field book stuck into his waistband against the small of his back. He was efficient and to the point, and became known for making decisions fast and keeping the discussion on track; in one hour he could cover a lot of ground. Moreover, rather than getting worn down by the responsibilities, he seemed to be thriving on the stress, like a runner gaining confidence as he progressed. Part of that no doubt was purely physiological. Burton had found quarters nearby, and he was getting some sleep now, and growing accustomed to the rhythms of working eighteen hours a day. More significant, however, was the luck of suddenly finding himself in a starring role.

Burton had grown up in modest circumstances in a suburb called New City, upriver from the Bronx. His mother was a nurse. His father was a New York precinct cop, who after retirement became a security guard at a suburban hotel—a job he continued to hold now. Burton had one brother and two sisters. As a boy he was unremarkable, an uninspired student lost somewhere in the middle of his class, and so unmotivated that he did not consider going to college until his parents told him late in high school that he would. He went off to Manhattan College, a Catholic school in the Bronx, because it was nearby, and he chose engineering there because he was never much of a reader. His roommate, Bill Cote, had a wider, more promising intellect, and they had good times together, but Burton seemed destined for an unexceptional life. Then somehow that changed: Burton's spirit did not necessarily widen in college, but he became alert to the possibilities of wealth and success. By graduation he wanted more than an engineering job could give him, and he went to work in the New York construction industry, and eventually earned an M.B.A. by taking night classes at a branch of Fordham University, in Manhattan. He worked hard, and began to make a good salary. It was the classic American story of a man's pulling himself up through sheer determination. When he went to work for the DDC, in 1996, it was not to take refuge, as so many do in government, but to gain experience in wielding power. He married a smart woman with a flourishing career in the telecommunications industry, and they had a child. By the time the Twin

Towers fell, they were expecting a second child, and living in the wealthy Westchester community of Chappaqua, near the Clintons, in a beautifully remodeled schoolhouse—a showcase of elegant living that Burton had created largely with his own hands. They had a live-in nanny and a vacation house at a ski resort in Vermont. They chartered sailboats in the Caribbean and Greece. They scuba-dived, played golf, and toyed with the idea of learning to fly. They threw catered parties, and sipped fine wine. It was not always easy to have to answer to Ken Holden, a bureaucrat who lacked Burton's technical training and shared neither his values nor his drive. But that was a temporary condition, which Burton could soon leave behind. He was a confident man. He believed that every move he had made so far had been right. And if by his own measure the success he enjoyed still seemed a little thin, he was certainly off to a good start.

Now came this attack. Burton's reactions in the first few days were patriotic, compassionate, and completely unselfish; he had not the slightest thought of advancing his career. Nonetheless, by the end of the first week it was impossible to ignore that a great opportunity had arisen here. People already were calling him a "czar," and the press had picked up on the metaphor, and at the twice-daily meetings he was wielding more power than he had thought possible before. In immediate terms the experience was exhilarating. Burton had a hard time not showing it. He gave some self-aggrandizing interviews to the press, most notoriously to the premier heavy-construction journal, a publication called *Engineering News-Record*. Holden thought the resulting article implied that the City of New York had hired Bill Cote and selected the four main contractors essentially because they were Burton's friends. Holden called Burton into an empty kindergarten room and threatened to fire him if he ever gave such an interview again. Holden pointed out that there were other people who could handle the job. Burton acted contrite, and may truly have appreciated Holden's warning. He was not just a runner with stamina to spare. He was a mountain climber with a route in sight.

Holden was something very different, and much harder to categorize. He was a generalist, a reader, the product of a liberal life. He grew up in a big house in Connecticut with several brothers and sisters, and a father who was a doctor and a mother who was a nurse. They were relatively religious, and regularly invited the needy from their synagogue to share their meals. They contributed to charities. They contributed to the Democratic Party. They had a lot of books, some philosophical, many about the Holocaust. Upstairs in the house the children were allowed to scribble graffiti on the walls. Holden's father preferred almost any activity to watching TV. He called *Star Trek* "*Star Drek*," and limited the kids to one hour of TV a week. He liked to play with words. He could finish the *New York Times* crossword puzzle in five minutes flat. He had a weakness for sayings like "Blackberries are red when they're green." When Holden was six, he had no idea what this was supposed to mean. When he was eight, his father began to make him do fractions in his head. When he was ten, his father started giving him essay assignments on such subjects as "Jams, Jellies, and Preserves" "Pirates, Privateers, and Buccaneers," and "Cornwallis's Surrender at Yorktown." These punishments fell on Holden through no fault of his own. When his parents vacationed in the Caribbean, he had to brief them with "Jamaica." When the neighbor's house burned down, he penned the classic "Spontaneous Combustion."

Holden escaped to a Jewish summer camp, which he enjoyed, and where as a teenager, at the end of his final year, he was caught in bed with a girl. To avoid expulsion he had to write yet another essay, this one on self-reform, which of course he did not take seriously. His father for once was openly proud. Holden went off to Israel for a year, came back, graduated from high school, went to Columbia, floundered, worked as a bike messenger, went back to Israel, where he farmed and carried a gun, and finally returned to Columbia, ready to study. He earned a degree in history, and for several years afterward lived the bohemian life in New York, briefly working for Columbia's alumni association and controller's office and then going deeper into the city, as the manager of a SoHo print shop. He was anti-establishment. He despised the military-industrial complex. He became a quasi-punk, and by the mid-eighties wore six earrings and had styled his hair into a modified Mohawk, dyed black. For additional effect he paid a tattoo artist on Long Island to engrave an image from the first Sex Pistols single onto his back. It was of two buses, one going to "Nowhere," the other to "Boredom." Years later, when his son would ask what the tattoo was about, it was hard to explain: this was from Daddy's nihilistic period, in the pre-dawn of history, not long ago.

In 1986 Holden was fired from the print shop, unfairly of course, and he spent six rough months on unemployment, doing a lot of reading. When the payments ran out, he snagged a low-level administrative position with the city's Parks Department. Then something unexpected happened. He began to put in long hours and apply the considerable power of his mind, learning such esoterica as the budgeting process, and producing top-quality work. Within the city's defiantly lethargic bureaucracy he soon stood out. Holden's co-workers demanded that he slack off, and he refused. After a year, the brilliant and eccentric parks commissioner, Henry Stern, sent word that he wanted Holden to clean up his image: the message was, he wanted Holden's hair "in a box." Meanwhile, Holden had gotten involved temporarily with an alluring secretary, a woman from the wild side of life, who wore red boots and ended up working in a topless bar, and whose most recent husband, a biker, began threatening Holden with violence. At last Holden decided he was ready for a change in direction. He did not send Henry Stern his hair, but he cut it off, and dropped his earrings, and began his rapid rise. Such was the unlikely but necessary formation of the man who emerged from the Trade Center chaos to guide the recovery effort through wildly emotional times, and to keep the equally necessary ambitions of Mike Burton in line.

The two men could never have been friends, but at PS 89 it didn't much matter: they had a job to get done. Holden did not attend the twice-daily meetings but stayed in the background, across the hall in the headquarters room, where he worked the cell phones incessantly. The goal was narrowly defined at first—to bring in equipment and move out debris in order to help with the search. There was no golden moment in which Holden and Burton were placed definitively in charge. Rather, there was a shift of power in their direction that was never quite formalized and, indeed, was unjustified by bureaucratic logic or political considerations. The city's official and secret emergency plans, written before the attack, called for the Department of Sanitation to clean up after a building collapse. A woman involved in writing the latest versions—a mid-level official in the OEM—mentioned to one of the contractors a week after the Trade Center collapse that she still did not quite know what the DDC was. This may explain her attitude when she found her way to the headquarters room on the second floor of PS 89, where Holden and Burton were talking one afternoon during the first week. She came up to them and said, "Who told you to get involved?"

Holden looked at her in disbelief. He was exhausted. He said, "We're kind of busy now. Why don't you come back in six months and ask that question." Holden and Burton were having a particularly difficult time because of the growing suspicion that the ruins now contained only the dead. That possibility in no sense eased the urgency. The pile out there remained so vast and wild that no one could quite yet give up hope that in some undiscovered pocket people might still be trapped alive.

That hope remained, if only as a glimmer, when, three weeks into the effort, the south face of the foundation hole's huge and protective "slurry wall" began to bow inward under soil and groundwater pressure from the outside. The wall was a reinforced concrete shell—eighty feet deep from street level to the bottom of its bedrock socket, and 3,500 feet around—that enclosed the Trade Center's ten-acre foundation hole and for thirty years had kept it from being inundated by the tidal waters that permeated the surrounding ground. It had been built by an injection technique, in which a slushy bentonite slurry was used to stabilize a deep trench (essentially a mold) before a steel lattice was lowered into it and the bentonite was displaced by a heavier mixture of concrete. The result was a buried wall whose inner face was exposed one level at a time, and then drilled through and stabilized with steel tentacles, known as "tiebacks," that were anchored into bedrock under the nearby streets and buildings. The tiebacks were temporary devices, subject to corrosion and designed to keep the slurry wall in place only during the years of construction. After the foundation hole was completely dug, the six levels of basement structure that rose within it served to brace the slurry wall from the inside, and, as planned, the tiebacks were cut and abandoned. All this elaboration was necessary because of the unnatural permeability of the Trade Center site: although the surface looked as solid as the rest of Manhattan, the land there consisted mostly of loose Colonial-era fill (garbage, shipwrecks, rubble, and so forth) through which New York Harbor and the Hudson River continued to seep, rising at high tide to within five feet of West Street. In essence the Twin Towers had been built inside a rectangular cofferdam—and now that dam was in danger of breaking.

The consequences of a break were disturbing to contemplate. In the worst case tidal waters would flood into the foundation hole so fast that even the largest emergency pumps would be overwhelmed, and any survivors still trapped on the lowest levels would be drowned. Moreover, the waters would gush uncontrollably through the entrances to the twin PATH commuter-rail tubes—cast-iron tunnels that penetrated the slurry wall near the lowest (B-6) level, and sloped under the Hudson to a slightly lower station in New Jersey, at Exchange Place. Such a flood would perhaps not be as destructive as the cataclysmic surge that had been feared on September 11, from a rupture below the Hudson of the tubes themselves, but it would likely have the same sequential effect, spreading through the New Jersey rail connections and back into New York around Greenwich Village, pouring into the West Side subways and causing unimaginable havoc with the functioning of the city. Spurred into action by this very real possibility, the Port Authority was hurrying to seal the tubes with concrete plugs. The plugs were to be built with man-sized portals, giving access to the tubes but designed to be slammed shut in the case of inundation. When the slurry wall began to move, however, the work was still days from completion. For the Trade Center site itself there were other concerns as well. If the Port Authority did manage to plug the PATH tubes, and the slurry wall subsequently failed, the water would have nowhere to drain, and the foundation hole would fill up nearly to the level of the street—an immense and debris-choked bathtub seventy feet deep. Because of the chaos and instability of the pile, the effects of such an inundation were impossible to predict, but they would certainly include enormous new complications and risks, and would mean at the least that the recovery effort would end up being measured in years instead of months. Three weeks after the collapses, therefore, the sudden movement of the south slurry wall posed an urgent new concern.

It was not, however, a surprise. Indeed, Richard Tomasetti had begun to worry about the slurry wall immediately on the evening of September 11, during the first walk-through with Burton and Holden, and that night he had enlisted the help of another renowned engineer, an underground specialist (known proudly as a "mole") named George Tamaro, who decades before had helped to oversee the construction of the slurry wall for the Port Authority. Tamaro was thought to understand the technology better than anyone else alive. He had prospered, and by now was the senior partner of the New York foundation firm Mueser Rutledge. At age sixty-four, he was a balding, avuncular man, urbane and charming, and as rumpled and self-deprecating as Tomasetti was dapper and proud. The two engineers entered the site together on September 12, and Tamaro, after surveying the ruins, quickly sketched a cartoonish map of the Trade Center's underground vulnerabilities for immediate distribution to the rescue operation—an outline of the slurry wall and also the location of four six-foot-diameter cooling intakes to the chiller plant, through which the Hudson would surge at high tide until the valves could be found and closed.

The destruction was of course impressive to Tamaro, who

commented with his usual verve that Dante himself would have been unable to describe such a place. Miraculously, however, the slurry wall seemed to have held. The evidence for this was indirect, because conditions on the pile precluded inspections, but it was persuasive. One day after the terrorist attack the PATH tubes were serving as massive drains for the foundation hole, with New Jersey (as usual) on the receiving end. The north tube had completely flooded at the mid-river low point, and both tubes were flowing heavily with brackish water. The Port Authority had set up powerful emergency pumps at Exchange Place and was extracting up to 6,000 gallons a minute to keep the station from flooding. Nonetheless, on the scale of possibilities this was still pretty small stuff—a flow easily accounted for by the surges through the cooling intakes, minor leaks in the slurry wall, watermain breaks, and the multiple streams of river water (much of it pumped by Sam Melisi's boat) that were being directed against the fires. In relative terms, therefore, the slurry wall remained watertight.

Tamaro did not celebrate, in part because he realized that the wall's survival was by no means assured. Indeed, it seemed intuitively obvious that the reserves of luck had already been used up. The Twin Towers happened to have fallen with incredibly tight vertical focus, each floor on top of the one below, a pattern that had lessened the direct hits against the wall. More important, while demolishing the basement structures that normally braced the wall against the outside water pressure, the catastrophe had crammed the foundation hole with two times 110 floors of debris—a chaos of steel and concrete that coincidentally also now worked to brace the wall. But that was the extent of the good news. As a bracing system, the rubble was frighteningly arbitrary, and judging from the constant collapses within the pile, it was also unstable. Now heavy equipment was heading toward the site, including huge self-propelled cranes (some, at 1,000 tons, the biggest ever seen in New York) that even without a shift in the debris could trigger a failure of the slurry wall, merely by burdening the soil nearby.

Tamaro was worried for another reason, too. For all the unexpected support provided by the debris pile, there was an exception on the south side of the foundation hole, and it was huge—a yawning crater sixty feet deep that had been blown through the basement structures by the collapsing South Tower, leaving ledges of broken concrete slabs hanging from the slurry wall, which for most of its length along Liberty Street now stood completely unbraced. Every calculation of the pressures that were acting on the south wall indicated that it should have fallen inward soon after the South Tower's collapse—and it will forever be a mystery why it did not. Convinced that the wall sooner or later simply had to fail, Tamaro waited impatiently for nearly two weeks while access was cleared, until finally he was able to start into a two-sided plan of salvation. First, a string of "dewatering" wells would be drilled through the ruins of Liberty Street parallel to the slurry wall to reduce pressure from the outside. Second, the hard-driving crew from Tully would attack along the wall's front face with hundreds of truckloads of dirt, which they would dump and bulldoze into the crater in an emergency "backfill" operation to buttress the wall from the inside. Tamaro knew he was involved in a race against time, and he was not at all confident that it could be won.

Nearly as great a concern was the west wall, which, though better braced by the ruins inside the foundation hole, was considered to be especially critical because of its proximity to the open waters of the Hudson River. A major failure there (whether spontaneous or triggered by a break of the vulnerable south wall) could produce the sort of immediate deluge that would threaten the workers on the pile. Once the backfill operation was under way, Tamaro decided to send an engineering team through an unexplored area of the underground in the ruins below the Marriott hotel, and along the west wall just north of the crater. Its purpose was to check the integrity of the structure, and particularly to look for heavy leaks, the existence of which could serve as an early warning of collapse. The team included, among others, the laconic fireman John O'Connell, as usual with cigar, and also Tamaro's duo of front-line engineers—the trim Andrew Pontecorvo and his giant partner, Pablo Lopez. I went along too, following Lopez through the tight spots. We climbed down through a hole in the surface of the pile, descended a ruined stairway beneath the hotel, and for several hours worked carefully through a dim and broken labyrinth, retreating occasionally to the filtered daylight along the jagged inner edges of the crater before returning into the darkness and continuing to probe through collapsed passageways toward the west wall. Lopez was the man who finally got through, half a floor up from the black floodwaters at the bottom of the foundation hole. He found the slurry wall on the far side of a twisted steel fire door, in a long, narrow space full of the sound of falling water. We edged along it, playing flashlight beams into the streams that were cascading down from overhead and squirting under pressure through the rusting old tieback connections.

Lopez said, "This is nothing. What we're looking for is Niagara Falls." He used a variety of terms for the leaks, mostly dismissive. Some he called "drips," some "pissers," and some, with just the beginning of admiration, "real criers." On that trip he found only one leak worthy of a compliment. It arced from the

wall about forty feet below river level, at a rate that Lopez estimated to be 100 gallons a minute. He called it "a genuine
gusher." It did not, however, qualify as an indication of real trouble on the west wall. When later that day he got a worried radio
call about a new leak on the exposed south wall, he diagnosed it
as just another pisser, and didn't even bother to go see it. He radioed, "Here, take an aspirin, and call me in the morning if it still
hurts." Judging from the silence on the other end, the message
was clear.

Lopez just naturally thought big. At one point, as he super-
vised the punching of a "mouse hole" through a ruined slab, I
mentioned that his "mouse" was about the size of a truck. He
said, "Bro! This is New York! So it's a rat hole!" He was thirty-
eight years old. Sometimes he wore a moustache, sometimes the
scrub of a beard. He'd had two wives and two children, and he
lived where he had grown up, in a tough part of Washington
Heights, in Upper Manhattan. His father had been a teamster,
his mother a test-tube cleaner in a lab. He attended Catholic
schools, and was an altar boy, but lost interest in religion. When
he was in high school, he wanted to get a part-time job. His
mother, who was the strong one in the family, said, "You stick to
your studies. The minute you get a taste of money, you're going
to want to buy a car. After that there is no end." Lopez eventually
learned to drive, but he never did see the need to buy a car. He
was such a city boy that when he rode the subway to college in
the Bronx, he thought it was like taking a train to the country.
The destination was Manhattan College, the same institution attended by Peter Rinaldi, Mike Burton, Bill Cote, Richard Tomasetti, George Tamaro, and, for that matter, Rudolph Giuliani.
There were so many Manhattan College graduates at the Trade
Center site that people referred to the work there as a school

project, and someone posted an alumni sign-in sheet outside one of the kindergarten rooms at PS 89, which rapidly filled up. Pablo Lopez studied at Manhattan College on a full scholarship. He told me he was given the scholarship because he had applied to study engineering and the admissions people figured they wouldn't have to pay out much money, because no Ecuadorian from Washington Heights would be able to maintain the required grades. If so, they were wrong. Lopez eventually earned a master's degree from Manhattan College, and taught there, too. He liked to joke that he had played the system for a fool. He was proud of his humble background, and his connection to the streets.

One day at the Trade Center site he went for dinner at the Red Cross feeding station, and he had salmon for the first time in his life. I asked him later if he had liked it, and he said sure, it was fine. He was not a fussy eater. The next day he returned to the feeding station, and for lack of choice had salmon again. He wolfed the fish down without giving it much thought. On the third day, after emerging from the underground, he and I went to the feeding station together—and again the offering was salmon, only stuffed. Lopez decided to speak up. The woman behind the counter was a Red Cross volunteer. Lopez said to her, "Hey, don't you guys have anything like hamburgers?" His tone was affable.

She smiled at him. "Don't you know salmon is good for you?"

"Yeah, but I'm a poor guy. I'm used to ground beef."

He was unshaven. He had a hardhat and a respirator. His hands were washed (a requirement for getting in), but otherwise he was streaked with the Trade Center's stains—its mud and ever present dust.

She kept smiling, and said, "I'll put in a special order."

Lopez half believed her. The Red Cross volunteers were unusually gentle with people in the food lines—presumably because they had been told that conditions on the pile were traumatic, and they believed that the workers required comfort. But when Lopez went back on the fourth day, the choice again was salmon. He'd had enough. On the fifth day he left the Trade Center and strolled down Chambers Street to eat dinner at McDonald's. He said it was the best meal he'd had in ages. He did not mean to exaggerate. We had been three weeks at the site, but it felt much longer.

And then, early one morning before dawn, under the harsh white illumination of emergency lighting, an ominous crack spread through the ground along Liberty Street, because the south slurry wall had begun to fail. During the days that followed, the Trade Center site teetered on the edge of a second great disaster. Peter Rinaldi was the first to report the trouble to Mike Burton. The crack lay thirty feet south of the wall, and was two inches wide. Though it didn't look like much yet, Rinaldi believed it meant the worst. He was extremely calm about this, in the style of a grown man talking about problems with his car. Burton, too, was calm—though obviously very concerned. At the morning meeting he said, "It may not mean anything, but the something it could mean would not be a good thing." People laughed. There was other business to take care of too. For instance: there had been a near fatality the night before on the pile, when a cable snapped and dropped a sixteen-ton beam into a hole that a fireman had just climbed out of—so people needed to be more careful about suspended loads. There had been a fire in Bankers Trust, ignited by the sparks of a cutting torch—so firemen would have to accompany the cutters there now. A piece of landing gear had been found on the roof of 90 West Street, and the owner wanted it off—but the FBI had not released it yet, so there was no point in trying to retrieve it. There was dangerous mold in the ruins of the Borders bookstore—so people needed to wear their respirators if they went inside. The EPA was insisting that workers had to use the new boot wash-down stations before leaving the site—so this would have to be done. The FBI, on the other hand, was insisting that the loads on the departing trucks could not be washed down, because they might contain "evidence"—but this seemed so out of touch that certainly for now the directive could be ignored. And finally, the Department of Health was demanding an effort to fight the growing population of rodents at the site—so who here had always wanted to be a rat king? Again, people laughed. The slurry wall was failing, and Burton was exhibiting the appropriate Trade Center aplomb.

Afterward Burton donned a hardhat and went with Rinaldi directly to the south wall. The crack was lengthening and widening; it seemed to be deep, too, because with a flashlight you could not see to the bottom. Rinaldi ventured that the crack represented a "classic shear failure," the back of a wedge of soil that was dropping and filling the space created as the top of the wall crept inward. The movement was confirmed by survey points—and it amounted so far to about four inches. At the first sign of trouble earlier that morning Rinaldi had briefly stopped the backfill operation in the crater, worried that it might be inflicting rotational forces, inducing the failure by pushing the base of the wall outward; but after rethinking the problem he had decided this was unlikely and had asked that the operation continue. He had also considered the possibility that the drilling under way for the dewatering wells, which paradoxically required the use of water around the drill bit, might actually have recharged the groundwater in the area, increasing the pressure on the wall. Burton himself did not bother to theorize. He immediately ordered all equipment, including the drilling rigs, away from the wall, and asked the dewatering contractor to activate the five wells already in place, despite the lack of a permit. Then he ordered an all-out acceleration of the backfill operation.

The days passed in a blur of roaring machinery and impromptu meetings in the smoke and dust of the pile. Tully rushed in truckload after truckload of fresh dirt, and bulldozed it hurriedly against the wall. But the wall continued its glacial advance. It was inexorable. Again suspicion arose that the backfill itself was causing the trouble—not by inducing rotation, but more simply by burdening the wall's attached ledges and slowly dragging the structure down. George Tamaro visited frequently to ponder the scene. On the third day of the crisis, with no improvement in sight, the key figures assembled once again on the ruins below the wall. Burton said, "George, what assurance do you have that the fill is not causing the collapse?"

Tamaro said, "Why don't you ask me when I'm going to die?"

Rinaldi said, "The problem is, the wall shouldn't even be standing."

Tomasetti was there too, well dressed as usual in a suit and good shoes. He said, "The question is, why is the collapse happening so slowly? The fill does seem to be having some positive effects." But he didn't seem at all sure of that.

By this time the top of the wall had moved six inches and was continuing to drift. In New Jersey the Port Authority had just completed the PATH tube plugs and now had the option, if necessary, of closing the Trade Center's drain. Tamaro thought it might have to be done. He said, "I don't want to be an alarmist, but . . ."

Burton said, "We know from the PA there's no increase in the outflow." This meant there was no increase in the inflow, either.

Tamaro said, "Yet."

Indeed, the pumping was accelerating in New Jersey, and the water was turning more brackish, as it would if the slurry wall were failing, allowing in the tidal waters—but word of this took a while to arrive.

Tomasetti said, "Can't we anchor the wall?"

Tamaro answered, "No. How? Struts? Tied off to what?"

Rinaldi said, "There's no clean, easy solution." His acceptance of this was a measure of the progress he had made.

The earth shuddered underfoot, as structures collapsed far below. Burton did not allow it to distract him. He assumed a position that became familiar through the worst of the crisis, standing by the drifting wall with his arms folded across his narrow chest and a frown of concentration on his face. His reputation was on the line: if the wall collapsed and later it turned out that the backfill operation itself was the culprit, he would be pilloried for years to come. Word arrived that people at City Hall were requiring the sort of assurances that no one could give them. Holden was running defense, trying to steady their nerves. Privately Holden was pessimistic. His reputation, too, was on the line. To me he said, "We're caught with our finger in a dike." But he was too shrewd a manager to second-guess Burton now. Here was a case in which Burton's faith in his own decisions might work to everyone's favor. He did not waffle as others did. He just said, "Keep the backfill going." He was very tough. People thought he was secure with himself, but that was not quite it. He had a climber's need to prevail. And this time, once again, he turned out to be right. There was no great skill involved. The dirt kept piling higher. On the fourth day the movement of the slurry wall began to slow, and on the fifth, having drifted an incredible twelve inches without failing, the wall finally stopped. That evening, Burton allowed himself a beer.

B y then a month had gone by since the attack, and twenty-eight days had passed without a single victim's being found alive. The sense of urgency remained strong, but people knew what they began privately to express—that the effort now would amount only to a search for the dead. It was going to be sad, frustrating work, and all the more disheartening because it was obvious that many of the victims had been obliterated without a trace, atomized by the collapses or cremated in the infernos that ensued. Indeed, by the end of the unbuilding and recovery effort, in summer 2002, fewer than half of the people presumed killed had been found and identified—and many of those only through the most ambitious program of DNA matching ever attempted in the United States.

Nothing about that search was easy. Even when they were whole and fresh, the corpses did not wait tidily on the surface to be collected. They lay crumpled and deeply intertwined with the rubble. Near the top of the pile, where the compaction was less severe, intact bodies were uncovered by diesel grapplers lifting heavy beams, only to prove difficult to extricate from the tangled debris. Firemen formed most of the recovery teams, and they directed the procedures. Affected no doubt by the isolation of the site as much as by their grief, they treated their own dead with a reverence not afforded others. Because they controlled the emotion-laden retrieval process for the entire site, their attitude bred factionalism on the pile, and in October led to an argument over the body of a Port Authority policeman that foreshadowed more serious confrontations to come. The policeman was discovered in the ruins of the Trade Center's plaza, and although his body was intact, one leg was pinned under a chaos of heavy steel that would obviously take hours to excavate. Because of the instability of the ruins, the excavation would also require shutting down other recovery operations in a wide area at the center of the site. The firemen gathering around had a better idea: they would free the body instead by cutting off the leg. Amputation may actually have been the right decision to make, but it was seen by the Port Authority police officers who were arriving on the scene as a solution based on the firemen's relative disregard for non-firemen, and their desire at all costs to keep searching for their own. This was probably unfair. But even when the police pointed out—correctly—that no dead fireman would have had his leg cut off, the firemen were unwilling to back down. The dispute never got physical, but it took a while to resolve, and it was not forgotten. In the end the Port Authority police won, and the center of the site was shut down for eight hours while they extracted their man. The shutdown became an act of tribute in itself. Because it was imposed, it also became an exercise in power. Increasingly within the inner world of the Trade Center site the dead were seen as members of different tribes.

The nature of the search for human remains necessarily underwent changes as the geography of the pile itself did. But a basic template for the process was established early. Normally there were seventy-five firemen on recovery duty at a time, supplemented by equal numbers of police officers from both the city and the Port Authority. They came in on one-month tours, and worked twelve-hour shifts according to a schedule that gave them every third day off, resulting in a pattern by which each searcher worked a total of twenty days. Though a few firemen signed on for multiple tours, most did not. This created a visible cycle at the site, between the overeager and incautious searching by teams at the start of their tours and the calmer, more efficient work that was performed later. Sam Melisi was one of the few who were present the entire time, on a relentless schedule of fourteen hours a day, six days a week, and in his self-effacing way he helped to steady the newcomers and teach them the necessary skills.

It was not complicated work, but it was tedious and grim, and because of the heavy equipment and unstable conditions on the pile, it was unsafe. For the most part what it required was patience. The searchers stood hour after hour in small groups by each of the diesel excavators that, one bite at a time, were tearing apart the ruins. They carried rakes and shovels, which they used to stir and probe the loose debris. As the operation matured, it settled into a two-stage inspection process, by which the searchers carefully examined each fresh load on the pile itself, and then took a second look at a debris-transfer point that was established at the southeast corner. The goal was to intercept remains at the earliest possible moment, on the pile in order to keep from scattering them and hindering the already difficult problem of identification, and at the debris-transfer point to spare the dead from what was seen as the indignity of being trucked and barged to the landfill at Fresh Kills—where, however, a third and particularly painstaking inspection was carried out.

The search imposed a reversal of normal sentiments about the dead, so the toughest days were those in which none were found. The firemen concentrated their efforts first on the debris where their colleagues were likely to lie, and as expected, they found pockets in the ruins of the stairwells in which some of the dead were stacked on top of one another. But the site never yielded the large concentration of victims—"the mother lode," Sam Melisi called it—that everyone was hoping for. Most of the dead were instead found "in dribs and drabs," as a discouraged fireman told me—meaning in ones and twos, and distributed all over the place. The best predictor turned out to be the nature of the materials being excavated—at one extreme, the intensely compacted and burned-through ruins that hardly merited a glance, and at the other, the relatively loose jumbles of heavy steel, where intact bodies continued to be found until nearly the end. I remember a startling moment late in December in the "donut hole" crater of Building Six, when a grappler lifted a beam and uncovered a man in a suit and tie who had fallen from the brokerage firm of Cantor Fitzgerald, in the North Tower, and was sitting upright, now somewhat shriveled but whole, with his wallet in his pocket. His condition was surprising particularly because he had been sitting almost in the open for a few months, and had not been entombed in the pumicelike Trade Center powder that helped to preserve others. Generally, the bodies that endured best were those of firemen, because they were wrapped in equipment and heavy clothing, whereas the most devastated were those of women, whose stockings and blouses offered poor protection during the collapses and after death. But, of course, decay began to set in for all of them from the first day, and the unseasonably warm winter weather sped the process along. By January the intact bodies being found were falling apart, with heads in particular becoming detached. At one of the morning meetings an official from the Medical Examiner's Office of the City of New York made a private plea to take greater care with the exhumations, because of the obvious undesirability of having to identify the same victim multiple times.

But most of the human remains were already in pieces; the force of the aircraft impacts and of the ensuing collapses had seen to that. Indeed, of the 1,209 victims identified within the first ten months of investigation by the Medical Examiner's Office, only 293 were nearly whole bodies, and the others were identified from among 19,693 parts. Of course, some of these parts were clustered, and some were collected at Fresh Kills, but the great majority of them were excavated, mapped, and tagged one by one in the chaos of the pile. Many of the parts found during the first few weeks were easy to distinguish. I saw bones scoured by fire, and a section of vertebrae with tissue hanging off, and a single severed arm with no sign of other remains nearby. It was the weight of the parts that made them seem most real. The arm was placed in the standard red biohazard bag, and it stretched the plastic as a fireman lugged it to an all-terrain "Gator," trying to hold it away from his body to keep it from bumping into his thigh.

After the first few weeks, work at the site settled into a certain rhythm. The fires continued to burn, water was poured onto the pile, the backfill operation delivered dirt to the south side to shore up the slurry wall, and the contractors began pushing in roads of finely ground asphalt, known as millings, which they laid on top of the rubble. All this and the passage of time meant that the scattered body parts began to look like lumpy cakes, similar to the ordinary debris into which they were mixed. The only way to spot them was to stand practically on top of the places where the grapplers were working. The placement of the night lighting was very important: without multiple angles of illumination shadows were created, and in those shadows entire lives could be missed. The safety officials at the site announced an exclusion zone of fifty feet around the machines, a rule so obviously impractical that people assumed it was merely for the record, something like, "Here are your orders (now please ignore them)." Sam Melisi said, "Fifty feet? You might as well be two hundred feet away with binoculars. You might as well not even go into the pile." He himself frequently reminded the other searchers about the dangers of working near these machines. But in practice it was usually a subtlety that gave the body parts away—a shred of clothing, the texture of striated muscle, a certain brownish color that was just a shade out of place in the gray-black mud of the pile. More important still, there was the smell—and you had to be close, with your respirator off, to pick it up. The human nose seemed to be correctly calibrated for the work on the pile, largely because of its insensitivity. Only once in my memory, in early November, was a large area enveloped in the stench of death. Otherwise the smell of rotting meat meant that a remnant lay practically underfoot, and with a rake or a shovel could quickly be found. Because of the Trade Center's restaurants and cafeterias, not every remnant was human. Word came from the forensic anthropologists working off the site that among the recovered materials were cuts of beef and chicken, and a significant number of hot dogs. This caused hardly a comment on the pile, where the difficulty of distinguishing one kind of meat from another was accepted matter-of-factly.

At the receiving end of the effort, in a building uptown known informally as the Dead House, sat a master of the matter-of-fact, the Medical Examiner's principal liaison to the Trade Center site. He was David Schomburg, a bald, middle-aged man who had been the butt of predictably macabre jokes of the "Igor" variety, and who indeed was pale, but who also possessed a lively intelligence and a necessarily unsqueamish approach to death. Though he rarely spoke at the meetings at PS 89, Schomburg turned out to be an impressive monologuist. I went to see him one winter day, and asked him about the normal functioning of the Medical Examiner's Office. He looked me in the eyes and began to talk. What he said, in shortened form, was "We are an independent finder of fact. Our duty is to take and conduct independent investigations of deaths under our jurisdiction. We serve all of New York City. We have jurisdiction over anyone who has a death arising out of casualty or injury. Trauma, for example. What you have to understand is there is a difference between *cause* of death, *mechanism* of death, and *manner* of death. Cause of death is the disease and/or injury that is responsible for the mechanism of death. Mechanism of death is the pathophysiologic means whereby the disease and/or injury exerts its lethality. Manner of death is a classification system—and it's like death is a multiple-choice event. We have six manners to choose from. One, we have natural death, which means it's caused exclusively by disease. Two, we have accidental death, which means it's a death arising out of an injury that came from an unplanned sequence of events: basically, we learned what 'accident' was the second time we spilled the milk at the breakfast table. 'Gee, Mom. It was an accident.' Three, we have homicide, which is death at the hands of another with degree of intent. Four, we have suicide, which is death at one's own hands with degree of intent; self-homicide, if you would. Five, we have therapeutic complication, which is a little complicated, but basically the easiest way to understand it is the 'but for' logic test, as in 'But for the therapy, the patient would be alive.' Six, we have undetermined, which is when we cannot decide the manner of death with reasonable medical certainty."

He took a breath. "Overall, our mission is to determine the

cause of death and certify the dead with their manner of death. I guess that's why people think the Medical Examiner's Office is just the Dead House. In one door, autopsy, and out you go with a certificate. But it's more complicated than that. There are parts of the death certificate that are opinion, and there are parts that are fact and legally binding. One of the legally binding parts is, who died? Where and when? It's our responsibility as certifiers to say that this is in fact the person who died. And we do that via a variety of ways. Sometimes it's easy. If you were to drop dead and people know you, they could look at you and say, 'Yeah, this is my friend Bill!' However, there are times when we have to realize that people are kind of like bananas. You buy a bunch of bananas, they sit on the counter, you eat a few, and there's always a couple that hang around, get a little freckled, and pretty soon they're not too appetizing. You stick them in the refrigerator to make banana bread later, and they get shoved to the back. Several months go by, you discover them back there, 'What the heck is this?' It's a rotted banana, and it looks like every other rotted banana. People are a lot like that. Left dead, unattended in warm environments, you can't tell them apart. Also, injury can take and disfigure or remove people's identifying features, make it very difficult. So we have to rely on other means—fingerprints, dental records, x-rays that were taken prior to death, and DNA analysis. It's embarrassing to have somebody you certify dead show up on the beach in the south of France enjoying the insurance money with a girlfriend. Or boyfriend. We routinely process twenty thousand deaths a year. But when it comes to IDs we generally don't rely on just one method. We're kind of fussy. That gets us to our involvement in the World Trade Center, which is interesting because our approach has been different. We felt early on it was unfair to make families wait years to have somebody declared dead, and that at some point we should be allowed to issue a death certificate even though we may not have been able to confirm the presence of a body. So we have actually two channels of certification going on. We have the real one, where we find somebody, identify who they are, and issue a death certificate. And we have this other side, which is a death certificate based on court declaration. And then when we do find somebody, we replace that court certificate. It's a long, tedious process, but the goal is still to identify as much of the material as we can and return it to families."

What Schomburg meant, in Dead House terms, was that he, too, was improvising in response to the Trade Center attacks, and pioneering new ground. We walked through the uptown processing facility to the end of the line, a large white tent where the remains were stored in twenty refrigerated trailers, and we stood there and talked. The discussion was clinical at first. It was about procedure, forensics, and physical conditions on the pile. For both of us it was all by that time rather routine. But then I asked him a question for which there was no simple answer: What is it in the bewilderment of loss that drives people to want so desperately to retrieve the bodies of their dead and, in the extremes of this case, to retrieve those bodies even after they are unrecognizable and torn to shreds? Schomburg was a specialist in death, and deeply involved in precisely that process, but even he was unsure. He said, "It's everywhere, I'd guess. The body becomes like the ultimate symbol. It represents the person we knew, the person we loved. And when that body no longer reacts—when it's dead—in our minds, now we have to accept that this is so. We've seen it with our own eyes." True enough.

But if that was the extent of it, if that was the only justification for this enormous effort of retrieval and identification, there was a legitimate question of how far a sane and healthy society should go along. Schomburg seemed to have anticipated my doubt. He said, "Or maybe it's just that every culture, every society, has its own kind of funeral rite."

The next afternoon it snowed, and by evening the site was transformed, covered by a tentative blanket of white, though muddy still along the access roads, and black amid the hot debris near the North Tower ruins, where the pile still smoked and vented steam. The crews had been anticipating the challenge of snow nearly from the start, and they spread salt to keep the trucks from slipping, and maintained the pace of the work without the slightest hesitation. The snow continued to fall, filling the air under the emergency lights with a density of crystals. It was a Saturday night, and out beyond the perimeter Manhattan was returning to a normal life. People here were enjoying the weather too, and profoundly, because it deepened the sense of inhabiting a special world. I ran into Sam Melisi on the pile near a recovery team, standing next to freshly turned debris; he looked like a northern woodsman with snow on his moustache. Pablo Lopez came over and complained about the cold as if it were an affront to his heritage. Late in the night Bill Cote and I walked the site; for a while we simply stood and gazed at the ruins. They seemed impossibly beautiful.

In the morning four inches covered the pile, smoothing its surface. I followed footpaths across the slippery steel, and broke a trail high into the ruins of Building Six for a better view. Below, the footpaths wove like little veins across the terrain, and joined the main access roads at the center, where the heavy equipment was hard at work. A thousand people must have been there overnight, tearing into the debris with their ferocious machines, but such was the scale of the scene that in the fresh snow now the ruins seemed hardly to have been visited. We'd come a distance nonetheless. The attack was still vivid in people's minds, but it was drifting into a category of memory assigned firmly to the long ago. I perched on the edge of a fallen steel column, and for a while did nothing more than let time pass. One of the grappler operators came up and sat beside me to smoke a cigarette. He was impressed by the overnight transformation, and said that it calmed him strangely. We spent a few words on it. There was something about the snow—its majestic indifference to human events—that seemed to provide perspective on what was happening here. Maybe it was just that we had become residents of such an intense private place. After so many weeks, whatever calm the snow could provide was welcome.

THE DANCE OF THE DINOSAURS

On the morning of Friday, November 2, 2001, seven weeks and three days after the Twin Towers collapsed, tribal fighting broke out at the World Trade Center site. The battle was brief and inconclusive. It occurred near the northwest corner of the ruins, when an emotionally charged demonstration turned violent, and firemen attacked the police. Within that intense inner world it came as no surprise. Resentments and jealousies among the various groups had been mounting for weeks, as the initial rush to find survivors had transmuted into a grim search for the dead, and as territoriality and the embrace of tragedy had crept in. The catalyst for the confrontation was a decision made several days earlier by Mayor Rudolph Giuliani to rein in the firemen, who for nearly two months had basked in overwhelming public sympathy and enjoyed unaccustomed influence and unlimited access to the site. In the interest of returning the city to normal life, Giuliani declared that the firemen now would have to participate in a joint command, with the New York and Port Authority police and the civilian heavy-construction managers in the DDC. This did not sit well. Moreover, access would be restricted, new procedures would be imposed at the pile, and the number of searchers would be reduced by two thirds.

The reason given publicly for this new arrangement was

"safety," a term so often used to mask other agendas in modern America that it caused an immediate, instinctive reaction of disbelief. Ordinary frontline firemen were the angriest. As many as 250 of their colleagues lay unaccounted for in the ruins, and they intended as a matter of honor to find every one of them. That week alone they had found fourteen. They were convinced that only they could sustain the necessary attention to detail on the pile, and that in his eagerness to "clean up" the site (a term they despised), Giuliani was willing to risk overlooking some of the dead, scooping them up with the steel and concrete and relying on the sorting process at Fresh Kills to separate their remains from the rubble.

To a degree the firemen's suspicions may have been well founded. Certainly the unbuilders themselves, operating under the direction of Mike Burton, were pursuing the most aggressive possible schedule of demolition and debris removal. Even Sam Melisi had doubts about the city's motivations. To me he said, "If you do a good job, and you do it in record time—who knows what record time is, since nobody's ever dealt with this before— is there a bonus? Do you get to be a commissioner or something? I don't know. Burton's just hard to figure out. Sometimes he's very personable and human: 'Oh, geez, you found X amount of people. Oh, that's great.' And then other times: 'You're really taking way too much time to look at this stuff, and we've got to keep moving.' Putting his arm around me and winking. 'We really have to move on this.' But you know what? We're going to take as long as it has to take. We're not going to compromise on that. We just can't do it. I don't have an allegiance to any construction company, or even to the City of New York. My only allegiance is to the people who lost their lives—to their families. The best we can do is try to retrieve as many people as we can in the most humane fashion. And then when all this is over, I can just go back to doing what I do."

Melisi was the reasonable one, and the most broadly involved of any person at the Trade Center site. By comparison, the ordinary firemen were narrowly focused on the rubble underfoot, where the remains of civilians and police officers were regularly discovered, but only the recovery of their own people seemed genuinely to interest them. Though their attitude was sometimes offensive to others working on the pile, it was not difficult to understand: the firemen were straightforward guys, initiates in a closed and fraternal society who lived and ate together at the station houses, and shared the drama of responding to emergencies. Some had lost family when the Trade Center fell, and nearly all had lost friends. Their bereavement was real. Still, for nearly two months they had let their collective emotions run unchecked, and they had been indulged and encouraged in this by society at large—the presumption being something like, "It helps to cry." The effect had turned out to be quite the opposite: rather than serving a cathartic purpose, the emotionalism seemed to have heightened the firemen's sense of righteousness and loss. Now, with the city ordering cutbacks in the firemen's presence on the pile, the agitation among the rank and file was so great that the firefighters' unions warned the city that they had lost control, and would have to organize a protest to avoid a break with their own membership.

None of this was conducive to clear, calm thought. Two days before the fighting broke out, a fire captain and union trustee named Matty James presented the situation starkly to the *Daily News*, as if retrieval of the bodies were an all-or-nothing affair. He said, "The city may be ready to turn this into a construction job, but we're not. We want our brothers back. By doing this, the city is taking away from these families, these widows, these mothers and fathers, any chance for closure. What are we supposed to do? Go to the Fresh Kills landfill to look for our people?" In a similar style, the mother of a dead fireman whose body had been found said, "Our memorial Mass became a funeral. It gave me an opportunity to hug the casket and to say good-bye. It's just awful, what they are trying to do . . . to deprive other women of that wonderful feeling."

Such was the rhetoric during the days leading up to that Friday morning, when hundreds of firemen began to assemble at the corner of Chambers and West Streets, by the red-brick walls of PS 89. In the emergency command center on the second floor, most of the site-management team from the DDC made plans to stay inside and out of sight. Mike Burton looked worriedly down at the scene in the street—at the slowly growing crowd, the opposing lines of blue-uniformed police officers, the television crews that were just then arriving. As if to himself he said, "These guys are not happy . . . not happy at all."

Earlier that morning Burton had led the first "combined" site-management meeting, at which the Fire Department, the police, and the DDC had been required to share the stage. It had taken place down the hallway, in the school auditorium. The various tribes had eyed one another with frank distrust, but had spoken in veiled terms to keep conflicts from breaking out into the open. Burton had said, "The DDC will have oversight [Meaning: control] over all recovery efforts on the pile," in a tone that somehow implied that this was a decision imposed on him from far above. He had also said, "We're making a push for increased safety. There may be some negative consequences. The police will not let anyone in or out of this building after ten A.M." (Meaning: The firemen are being hysterical.) A fire chief in a white uniform shirt seemed not to have heard a word. He said,

"We're trying to hammer out a plan for the removal of victims and uniformed personnel. We've just located fourteen Fire Department members near the South Tower." (Meaning: We're not going to make this easy.)

In the kindergarten room Holden decided that he needed to go to the pile to keep a close watch on the situation himself. Peter Rinaldi expressed concern for Holden's safety. Holden said, "I'm just going to wear a sign saying, 'I don't know Mike Burton.'" Burton smiled humorlessly. His relationship with Holden was increasingly strained. There was another Port Authority man in the room, a strapping ex-Marine named Tom O'Connor, whose wife had worked in a neighboring building. She had survived, but they had lost many friends in the attack. He said, "We are now viewed as monsters." His manner was matter-of-fact.

Becky Clough, a pugnacious DDC manager and one of the few women active on the ground at the heart of the site, said, "What are we afraid of? We're doing the right thing."

Burton said, "It had to happen sometime anyway. Maybe there are some positive sides."

O'Connor said, "So when are you losing your job?"

I slipped outside before the building was locked down, and walked through the demonstration. The crowd by then had grown to perhaps five hundred off-duty firemen, and it was continuing to expand as others streamed in from the subways and parking lots. Some were dressed in civilian clothes, but most wore the standard protective "turnout coats," black with yellow reflector strips and FDNY written in block letters across the back. Most were bareheaded. As protesters, they seemed awkward and self-conscious at first, and unsure of how to proceed, but they were also genuinely angry. Encouraged by a union official with a bullhorn, they raised their fists and chanted, "Bring our brothers home!" They waved a union banner and American flags. The TV crews moved in tightly, and found firemen for one-on-one interviews. A retired fire captain said, "My son Tommy is still in that building, and we haven't gotten to him yet." The firemen began shouting, "Bring Tommy home! Bring Tommy home!" The unions had promised City Hall that the demonstration would be orderly, and that it would remain outside the site's perimeter—that it would amount to little more than a show for the cameras. But the firemen proved difficult to control. As the crowd swelled to nearly a thousand, it grew louder and more confident, and suddenly surged south through the first police barricades toward the Trade Center ruins.

The TV crews followed eagerly. To the police the demonstrators shouted, "Walk with us!" and "Shut 'em down!" But the police were on duty, and after weeks of growing resentments on the pile they were not inclined to sympathize. As the demonstrators shoved through a second line of defense, the police shoved back, and some of the firemen started to swing. I saw one policeman go down with a roundhouse punch to the face; others responded, tackling and cuffing the offender. The crowd kept pushing through, with fights breaking out where the two groups met. These were big, physical guys on both sides, and they grappled in ungainly dances, straining hard, cursing each other, and toppling to the ground in aggressive embraces. Five policemen were injured. On the periphery the protesters who were arrested—twelve in all—were hustled into police vans. One was a tough-looking old man in a fireman's uniform, who kept bellowing, "My son's in there! My son's in there!" Firefighters shouted, "Let him go!" But like the others who had been arrested, the old man was hauled away.

The protesters gathered on West Street beside the ruins, where they were joined by a scattering of die-hard union sympathizers from the site—primarily a group of ironworkers who sauntered over out of curiosity, and got into the spirit of things. One in particular seemed to delight in showing the crowd how to go about making a TV appearance. He was the very image of a beefy construction worker, dressed in a hardhat and a soil-stained thermal undershirt. He climbed onto a diesel excavator and for several minutes mugged wordlessly for the cameras, waving an American flag and pumping his fist in the air. It was not clear that he knew or cared what the protest was about.

The same was obviously not true of the firefighters' union official, who climbed onto the excavator and used his bullhorn again, repeating his cry of "Bring our brothers home!" and threatening to find reinforcements by the thousands, including "brothers" from other cities, if Giuliani did not back down. Exhibiting more bravado than political sensitivity, he called on the police to release the men who had been arrested, and to march with the demonstrators on City Hall. But of course the police were angry at having been attacked—all the more so because they, too, had tribal allegiances, and had lost twenty-three colleagues in the Trade Center collapse. The firemen marched to City Hall, where they chanted, "Rudy must go!" This was somewhat gratuitous, since Giuliani was only two months from the end of his tenure. They also chanted for the ouster of their fire commissioner, Thomas Von Essen, who for years had been a well-liked New York fireman and union leader, but who now was going around quietly making the point that by far the greatest loss of life had been civilian, and that the Trade Center tragedy was larger than just a firefighters' or even a New Yorkers' affair.

Giuliani was infuriated by what he viewed as an assault on the city's all-important process of recovery, and he lashed out with the vindictiveness for which he was known. With the demonstration still going on, he called Holden on Holden's cell phone and demanded that he identify the ironworkers and fire them. This was only the second time in eight years that Holden had received a call from the mayor, and he was aghast at what he was hearing. Very few ironworkers had joined with the firemen, but those who had were likely to be union activists, and therefore just the sort of people who could rally sympathy across the pile. Any attempt to discipline them could easily backfire and lead to a full-scale rebellion. Moreover, for all he knew, the ironworkers were on their break (or could claim to be), and since they were not disrupting the work, they had every right to protest. This was the inner world of the Trade Center site, an emergency zone, yes, but not subject to martial law; if it was a turbulent and quarrelsome place, it was also courageous and creative, and an authentic piece of American ground. But Holden knew better than to argue with Giuliani, and he did not try. After ending the call, he dutifully had a DDC staffer at the scene take pictures of the ironworkers. It was a particularly unpleasant moment: the workers had no idea what the snapshots were about, and some posed for them festively, as if they were on a weekend outing. Holden looked disgusted by the whole affair. He was irritable in a way I had not seen before. As the demonstration drifted away from the pile, he came upon a New York tabloid-TV crew that was setting up to interview a man in an FDNY T-shirt—presumably about the depths of his sorrow. Holden told me he was tired of all the exploitation. He walked up to the reporter and demanded that he take his cameraman and leave. The reporter said, "We just want to . . ."

"Out!" Holden snapped, pointing north.

The reporter gave Holden a look of pure hatred. He said, "The man's got no heart."

"Out!"

By late afternoon Holden was sitting again in the kindergarten room of PS 89, thumbing glumly through pictures of the protesting ironworkers. He had no intention of following through with Giuliani's orders, but he knew that if he flaunted his disobedience, the repercussions would be swift and severe. He himself had told me earlier that the first lesson of "commissioner school" was "Don't contradict the mayor." The only alternative now was to procrastinate, and to hope that the idea would somehow go away. This was not the style of the Giuliani administration, in which the mayor's whims were treated as dictates. Indeed, during several phone calls that followed, one of Giuliani's deputy mayors, Tony Coles, continued to demand punishment for the ironworkers long after it could conceivably have served a useful purpose. But Holden held the line, in this case for inaction, and with defensive skills unheralded even within PS 89, he managed to protect Giuliani from himself, and the nation from Giuliani, and to keep the recovery effort on track.

It was a troubled time anyway, that first half of November, the low point in New York's response to the Trade Center attack. The Yankees had lost the World Series—and to Arizona, of all teams. The various groups at the Trade Center site were turning into warring camps. To make matters worse, Giuliani seemed to have lost control of his emotions. After the demonstration (soon known as "the firemen's riot") he continued for several days to rage about the protesters' distortions, and what he saw as their betrayal of the city's well-being. He did not seek conciliation—for instance, by forgiving those who were involved in the demonstration. Rather, the police hunted down another six firemen identified as culprits, and booked them on charges of criminal trespassing. Among these men were the presidents of both firefighters' unions: Captain Peter L. Gorman, of the 2,500-

member Uniformed Fire Officers Association, and Kevin E. Gallagher, the protester with the bullhorn, who headed the 9,000-member Uniformed Firefighters Association.

These were honest union officials, expressing the legitimate if misguided dissent of their membership, and their arrest was unusual, to say the least. It was also, of course, counterproductive. The firemen were outraged. Captain Gorman, who had worn the uniform for twenty-eight years, said, "They're putting me through the system like I'm a thug." He called the fire and police commissioners "Giuliani's goons," and Giuliani himself a "fascist." The unions threatened privately to hold a news conference and accuse the mayor of being "anti-American," but apparently thought better of going public with such a foolish claim. Instead, more accurately, a union spokesman said to *The New York Times*, "The mayor fails to realize that New York City is not a dictatorship, where if you don't like what a union is doing you can just go and lock up a union's president. The message being sent from City Hall is that if you don't agree with this administration, we will get you." Outside the Trade Center site America bloomed with bumper stickers proclaiming UNITED WE STAND, a strangely forlorn slogan in a country that so obviously draws strength from disagreement. Drawing strength from disagreement is a trick that the attacking terrorists must certainly have discounted. But of course there are also limits to the creative power of disunity—and the Trade Center response seemed to be veering toward just the sort of social implosion that the terrorists may have had in mind.

The tribalism that grew up on the pile had origins so primitive that they can only be understood as instinctual. At the core was an us-versus-them mentality brought on by the mere act of donning a uniform. Whether as firefighters or as the two sorts of police (city and Port Authority), the uniformed personnel at the site were generally drawn from the same white "ethnic" outer-borough neighborhoods and families, but as members of their respective organizations they had learned to distrust and resent the others. The hostility was historical, and because it was strongest on the lowest levels, among the rank and file, it had proved impossible to root out. People at the site referred to it alliteratively as the Battle of the Badges. Across the years it had led to frequent arguments over turf and occasional bouts of outright obstructionism at emergency scenes. At the Trade Center it had been a factor from the first moments after the attack, when the Police and Fire Departments had set up separate command posts several blocks apart, and without communication between them. There were consequences to this: after the South Tower fell, police helicopter pilots took a close-up look at the fire in the North Tower, and twenty-one minutes before the final collapse they urged their own command to evacuate the building. The warning was radioed to the policemen inside the North Tower, most of whom escaped, but it was not relayed to the fire commands, or to the firemen in the building, only some of whom were able to hear independently radioed orders to evacuate, and more than 120 of whom subsequently died. The lack of communication was certainly no more the fault of one side than of the other, but it aggravated the divisions between them. Even during the initial desperate search for survivors the police and firemen quarreled over turf, and asserted their differences. By the end of the first day the bucket brigades had separated according to uniform. Throughout the months that followed, individual friendships and family ties cut across the lines. Nonetheless, the tribalism festered and soon infected the construction crews, too, who did not quarrel much among themselves but generally distrusted the police as ordinary citizens do, and who probably hadn't given firemen much thought before, but came now to resent their claims to special privilege on the pile.

The firemen's claims were based on an unspoken tribal conceit: that the deaths of their own people were worthier than the deaths of others—and that they themselves, through association, were worthier too. This was difficult for the police and civilian workers at the site to accept. The collapse of the towers had been anything but certain. The firemen who had gone inside had been normally brave—as people are who are not cowards. They were not soldiers crossing the lip of a trench or assaulting a machine-gun nest in battle; they were men with a job that demanded mental willingness and hard physical labor, and on that day they were climbing endless stairwells one flight at a time in the company of friends, and with little obvious purpose in mind beyond finding the civilians who must have been injured by the twin attacks. The firemen in the South Tower were killed without warning. They were unintentional martyrs, noncombatants, typical casualties of war. Those in the North Tower felt the rumble of the South Tower's collapse, but as best as is known, most did not understand what had happened or conclude that their own building soon would come down. Later on, as the precariousness of the North Tower became clearer, there were firemen who committed acts of extraordinary heroism—for instance, by lingering to help civilians, or remaining in the lower lobbies, desperately working the radios and calling for an emergency retreat. But the Fire Department had no monopoly on altruism that day, and terrible though its casualties were, with 343 dead, it did not suffer the greatest losses. As the workers on the pile knew all too well, that sad distinction went to Cantor Fitzgerald, where 658 people had died—some of them no doubt as altruists too. But what did such categories mean anyway? Nearly 3,000 people had been slaughtered here, nearly all of them on the job, and each of them at the last instant equally alone. Those who had not been vaporized lay scattered in the rubble's democratic embrace. The dead were dead now and didn't care. And it was absurd for the living to group and rank them.

Were it not for all the hype, this would hardly need saying. It's true that the United States was shaken, and that people in their insecurity felt the need for heroes. The dead firemen certainly fit the bill. They were seen as brawny, square-jawed men, with young wives and children—perfectly tragic figures, unreliant on microchips or machines, who seemed to have sprung from the American earth like valiant heroes from a simpler time. They had answered the call of history, rushed to the defense of the homeland, and unhesitatingly given their lives. They had died at the hands of barbarians, leaving behind widows who were helpless, or who were said to be. All this presented opportunities for image-making that neither the media nor the political system could resist. Progressives may have been shocked by the ease with which America slipped into patterns of the past—with women at the hearth, men as their protectors, and swarthy strangers at the gate. Rationalists may have worried about the wallowing in victimization, and the financial precedents being set by promises of payouts to the victims' families. As usual in America, there was reason for everyone to deplore the cynicism and crassness of the press. However, there would be time for refinements later. As an initial reaction to the first shock of war, the hero worship was probably a healthy thing, as long as it was confined to the dead.

But when it spilled over indiscriminately to the living, problems arose, particularly at the center of attention, on the Trade Center pile. The firemen now on the scene were by definition those who either had escaped from the Twin Towers before the collapses or, more likely, had not been inside them to start with—in most cases because they were somewhere else in the city at the time. They were not lesser men for this. But if the loss of the others was to mean anything beyond the waste of war, it had to be admitted that people on the pile since then, though ferociously dedicated to a grim and dangerous task, were simply not involved in heroics. Of course the situation was presented differently on the outside, where the public was led to believe that conditions on the pile were so difficult that merely by working there people were sacrificing themselves, and that the firemen in particular—anonymous figures who wore the same wide-brimmed helmets as their fallen brethren—deserved the nation's adoration. For many of the firemen, who tended to have led quiet lives until then, the sudden popularity became a disorienting thing. Even those with the strength to resist the publicity—who stayed off TV, and did not strut in public—seemed nonetheless to be influenced by this new external idea of themselves as tragic characters on a national stage. The image of "heroes" seeped through their ranks like a low-grade narcotic. It did not intoxicate them, but it skewed their view.

Strangely enough, it was this patriotic imagery that ultimately drove the disunity on the pile, and that by early November nearly caused the recovery effort to fall apart. The mechanisms were complex. On the one hand, there were some among the construction workers and the police who grew unreasonably impatient with the firemen, and became overeager to repeat the obvious—in polite terms, that these so-called heroes were just ordinary men. On the other hand, the firemen seemed to become steadily more self-absorbed and isolated from the larger cleanup efforts under way. The resentments rarely erupted into fistfights (though fistfights did occur) but increasingly were expressed in private conversations on the pile—often on the subject of the looting that for the first few months tarnished the Trade Center response.

The looting was shadowy, widespread, and unsurprising. The Trade Center was known to have been hit before by errant policemen and firemen, after the terrorist bombing of 1993. This time the thievery was less intense but longer-lived. It involved small numbers of construction workers and men from the same uniformed groups as before, and it was shallow and opportunistic rather than deeply criminal in intent. It started in the shopping complex, with the innocuous filching of cigarettes and soda pop, and expanded into more ambitious acquisitions. As rumor had it, the tribalism at the site extended even to the choice of goods. Firemen were said to prefer watches from the Tourneau store, policemen to opt for kitchen appliances, and construction workers (who were at a disadvantage here) to enjoy picking through whatever leftovers they came upon—for instance, wine under the ruins of the Marriott hotel, and cases of contraband cigarettes that spilled from U.S. Customs vaults in the Building Six debris. No one, as far as I know, stole women's clothes, which hung on racks for months, or lifted books from the Borders bookstore, which were said to be contaminated with dangerous mold. After a few arrests were made, the filching shifted to the peripheral buildings, which were gradually thinned of computers until the authorities wised up and posted guards. It's important to realize that these transgressions occurred not in a normal part of the city but in a war zone, where standards had changed, food and supplies were provided free of charge, and a flood of donated goods (flashlights, gloves, Timberland boots) was believed to be backwashing onto the streets. It was also a place where the entire nation had been attacked and was responding as a collective, and where therefore, surprisingly for modern America, the meaning

of individual property had been diminished. In context the looting simply did not seem shocking.

Knowledge of it, however, cast a shadow on the use of the
word "hero," and at least once became a source of embarrass-
ment and bitter mockery. One autumn afternoon, at the base of
the South Tower ruins, diesel excavators were digging into unexplored reaches of the Trade Center's foundation hole. Fifty feet
below the level of the street they began to uncover the hulk of a
fire truck that had been driven deep by the collapse. The work
was being directed by the field superintendent for one of the major construction companies, a muscular and charismatic man who
was widely admired (and to some extent feared) for his un-
abashed physicality and his manner of plunging unhesitatingly
into battle with the debris. If for no other reason than his
confidence in the enormous mechanical power at his disposal, the
superintendent believed in acting first and worrying about the
consequences later. Early on he made it clear to me that were he
in charge, he would clean up the site in no time flat, and that his
first step would be to throw the firemen off the pile. Such was his
disdain that he might even have included Sam Melisi in the toss,
hard as that was to imagine. He assured me that he'd had nothing
against firemen before (he shrugged and said, "Why would I?"),
but he just couldn't stand this hero stuff anymore. He didn't like
the moralistic airs these guys were putting on. He didn't like the
way they treated the civilian dead. And he especially didn't like
the fact that they kept forcing his operation to shut down—once
for three days straight—while they worked by hand and poked
through the rubble for their colleagues' remains.

Imagine his delight, then, after the hulk of the fire truck appeared, that rather than containing bodies (which would have required decorum), its crew cab was filled with dozens of new pairs
of jeans from The Gap, a Trade Center store. When a grappler

pulled off the roof, the jeans were strewn about for all to see. It was exactly the sort of evidence the field superintendent had been waiting for. While a crowd of initially bewildered firemen looked on, the construction workers went wild. "Jeans! Look at these! Fucking guys! Jeans!" It was hard to avoid the conclusion that the looting had begun even before the first tower fell, and that while hundreds of doomed firemen had climbed through the wounded buildings, this particular crew had been engaged in something else entirely, of course without the slightest suspicion that the South Tower was about to hammer down. This was not what the firemen wanted to hear. An angry fire chief tried to shut the construction workers up. He offered an explanation—that the jeans (tagged, folded, stacked by size) had been blown into the crew cab by the force of the collapse. The field superintendent, seeming not to hear, asked the fire chief to repeat what he had said. When he did, the construction workers only jeered louder.

Scattered jeans lay on the pile for several days. The story got around. For Ken Holden and Mike Burton, this and other incidents on the pile amounted to important lessons in their war's early months: the site would never stand united, as sloganeers said it should, so some other approach would have to be found.

Though the firemen who rioted on November 2 did not believe it, when Giuliani gave "safety" as the reason for reducing their presence on the pile, he was completely sincere. This was somewhat counterintuitive, since the safety record so far had been extraordinarily good: despite the fires, the instability of the ruins, and the crushing weight of the equipment and debris, not a single recovery worker had been killed, and only a few had been seriously injured. Indeed, discounting possible long-term respiratory problems, the injury rate was about half that of the construction-industry average. Some people claimed this as a sign of God's favor, but a more mundane explanation was that the inapplicability of ordinary rules and procedures to such a chaotic environment required workers there to think for themselves, which they proved very capable of doing. Nonetheless, the city had reason to be especially concerned about the firemen at the site, who formed maverick groups on the pile, prone to clustering too close to the diesel grapplers and to taking impetuous risks in the smoke and debris.

The lack of discipline was a well-known aspect of the firemen's culture. In some ways it was a necessary thing, hard to separate from their views of manliness and bravery and their eagerness to take on fires. It also, however, led to needless danger. After he left the service, in January of 2002, Commissioner Von Essen mentioned that as a longtime firefighter, he had often stood on floors among thirty others where only ten were needed. Indeed, on September 11 many of the firefighters who responded were off duty at the time, and many bypassed check-in procedures, or arrived by subway or car in violation of orders to stay away. Many also went into the towers unnecessarily and with little coordination, at a time when there were enough other firemen on duty to handle the evacuations, and when the Fire Department had decided that the fires were unfightable. Sixty off-duty firefighters died that day. "Courage is not enough," Von Essen later said. "The fact that the guys are so dedicated comes back to hurt them down the line." The police at the site were better disciplined—and, partly as a result, they suffered fewer casualties on the day of the attack. But nearly two months after the tragedy, with no conceivable justification for continuing to jump into voids or clamber across unstable cliffs, there were still firemen running wild.

Giuliani had good reason, therefore, to rein them in. Viewed from the outside, the plan seemed sensible: you scale back the searchers to three teams of twenty-five—one from each of the uniformed services; you allow only one spotter at a time to stand beside each diesel excavator on the pile; until human remains are found, you require the other team members to wait in designated "safe areas" nearby; you do not allow ordinary firemen to keep shutting down the site; you create a joint command to soothe people's egos but give practical control to the engineers and the construction types, who are businesslike and know how to finish the job; you shrink the perimeter, with the goal of returning even heavily damaged West Street as soon as possible to the city; you thin the crowds of hangers-on by requiring new badges for the inner and outer zones, and asking the Red Cross and the Salvation Army to consolidate and simplify their feeding operations; you scale back the public displays of mourning; you encourage people to get on with their lives.

The planners were not completely naive about the transition: they suspected that they might have some difficulty in getting the firemen to comply. Mike Burton decided that the best approach would be to start with strict enforcement of the new rules, and loosen up later for the sake of efficiency. I was surprised by his confidence that anything here could go according to plan. And indeed, little did.

Questions of personality and professional formation were at play. The construction crews, like the DDC itself, were made up of hard-driving people, accustomed to shaving minutes in a time-obsessed industry. Though they understood the desirability of finding the human remains at the Trade Center site, they were not going to slow the excavation of the ruins just to ensure that the final inspections at the Fresh Kills landfill did not turn up body parts. Mike Burton in particular was pushing for speed, and was determined to finish the job below cost and ahead of schedule—however arbitrary those targets may have been. He was climbing his mountain of success, and was not about to let a gang of irrational firemen get in his way. In public and during the morning meetings he was gracious and respectful toward them, but in private—in the confines of the kindergarten rooms, or during long walks with me on the pile—he let his impatience show. On the evening after the riot we came upon one of the new search-team "safe areas" (a ten-foot square of Jersey barriers spray-painted FDNY), and he shot me a hard little smile of victory. He used the firemen's term for it and said, "The penalty box." I assumed he was thinking about the attack on the police. He may have believed that the way forward was clear. He seemed not to notice that the penalty box was empty.

This was not a game. There were no rules. The firemen continued with their headstrong ways on the pile, refusing to submit to civilian authority. Five days after the riot, after the unions formally apologized to the police, Giuliani began a partial retreat. He said he would increase the size of the search teams to fifty. The firemen were unmollified. The place where their friends had been killed was still being turned into an unholy "construction site." Three days later, on November 10, charges were dropped against all but one of the arrested demonstrators. It did no good. The firemen pulled out the stops and demanded a meeting between the mayor and the dead firemen's families.

The meeting took place behind closed doors in a Sheraton hotel in midtown Manhattan, on the evening of November 12. It had been a rough day already: that morning an American Airlines flight departing for the Dominican Republic had crashed into a residential neighborhood in Queens, killing 265 people, and most of the officials at the meeting had visited the scene. Now they sat behind a table on a raised platform—the mayor, the medical examiner, the fire commissioner, and, from the Trade Center site, Mike Burton and Bill Cote. The crowd they faced consisted mostly of widows—an increasingly organized group that spoke for mothers, fathers, and children as well, and that after two months of national sympathy was gaining significant political strength. Payouts to the victims' families hadn't yet begun, and firemen's widows, not without reason, felt neglected and put-upon. They believed that the city was essentially giving up on the search for the dead. And they were angry about it.

The medical examiner was the first to come under fire. He had begun to talk about the procedures in place for handling remains when he made the mistake of mentioning that full or even partially intact corpses had not recently been found, and that they were unlikely to be found in the future. A woman stood up and yelled, "You're a liar! We know what you're finding! You're a liar!" Others chimed in, shouting that their husbands' bodies had been recovered in good condition. One woman yelled that when her husband was found, the searchers on the pile could even see the dimple in his chin. It was as if an emotional dam had burst in the crowd. The medical examiner listened somberly. When the crowd briefly quieted, he tried to explain his reasoning: the excavation had moved into the mid-levels of the ruins, where the debris was severely compacted and the dead had been shattered or vaporized; furthermore, from what was known of the pile's composition, along with the processes of organic decay, there was little chance that whole bodies would be discovered in the future. The widows would have none of this. They continued to shout "Liar!" until the medical examiner sat down. The months ahead would show that the medical examiner was wrong——that the ruins were riddled with unexpected cavities deep down, and that nearly whole corpses, particularly of heavily clad firemen, lay waiting to be found. However, this would have been impossible to predict at the time.

The mayor handled himself well that night. He was patient and compassionate, and he allowed the grieving crowd to rail, but he did not pander to it. About the medical examiner he simply said, "He's not lying. He's telling you what we know." Then it was Burton's turn to talk.

Burton started gamely into an explanation of the transition on the pile, including the new placement of spotters, the "safe areas," and the handling and inspection of the debris. The crowd listened sullenly for a while, until a woman stood up and yelled, "We don't even want to hear from you! You're Mr. Scoop and Dump!"

Burton was flustered. He said, "Listen, this will only be a few more minutes. Just let me explain our thinking, so we're all on the same page and can have a rational conversation."

The woman shouted, "No! You're not the sort of person I want to talk to! You're the problem!"

Burton tried a soothing tone. He said, "We'll get through this, if you'll just . . . "

"No! You're Mr. Scoop and Dump!"

Others joined in, shouting, "Scoop and Dump! Scoop and Dump! Scoop and Dump!" Burton allowed a small, nervous smile to flicker across his face. His tormentor saw it. She yelled, "You're smirking? You're smirking at me? You think this is funny? This isn't funny!" Someone else shouted, "Yeah, he's smirking!" and again the whole crowd started in. Burton was mortified. He made a few weak attempts to speak, but the widows were relentless, and they overwhelmed him. For minutes he stood miserably on the platform, absorbing the abuse, unable to advance or retreat. Bill Cote felt terrible for him, and wanted to go to his rescue, but could not. Finally the mayor stepped in and got the crowd to simmer down. He told a little story and made a quip that caused Burton's tormentor to smile. The mayor noticed and said, "You see, you just smirked. When people are nervous, they smirk. So can we please put this behind us now? Let's just stop."

But the widows were too angry for that, and they soon widened their attacks. They had some legitimate complaints— for instance, that the Fire Department had never gotten around to contacting some women about their husbands' deaths, or to clarifying the associated administrative and financial details. A few of those whose husbands had yet to be identified were still having a hard time accepting their demises. And what about their paychecks? If a fireman remained trapped somewhere inside the pile, wouldn't he still be on duty and earning overtime? Conversely, was it correct to assume that all those who had disappeared had stopped working at the moment of collapse? Unexpected though these questions seemed, they were obviously practical, and the fire commissioner admitted that the department had done a poor job of handling such things.

For the most part, though, the widows simply vented their emotions. They argued as much among themselves as with the officials at the head of the room. One woman kept insisting that her husband was still alive, because she could send signals to his beeper and it would respond. Others, who had accepted the reality of death, were infuriated by the possibility that any of the firemen's remains would not be found until they reached the landfill. Not surprisingly, this turned out to be the most difficult issue of the night. The crowd demanded to know why the final sifting operation could not be moved from Fresh Kills to the Trade Center site or the streets nearby. There were many reasons why—including dust, noise, neighborhood opposition, and, most important, the complete lack of space in Lower Manhattan—but neither Burton nor Cote was about to say that now. They promised to look into the possibility. Of course the crowd did not believe them. At one point a woman came forward with her son, a boy of about seven, and started screaming, "You tell him! You tell him!" Burton and Cote looked at her without understanding. Tell him what? She continued, "You tell him his father's going to be found at the dump!" The crowd broke into applause. The woman began to cry. "Tell him! Tell him!" Her son watched her in apparent confusion. Friends came up, put their arms around her, and led her and the boy gently away.

Burton and Cote were badly shaken. When the meeting ended, after more than three hours of emotional storms, the two of them got into Burton's Jeep and drove away through the quiet streets. At first they did not speak, except briefly to agree that the experience had been the worst of their lives. In the theater district they found a bar, and went in for a drink. The other customers there—tourists pioneering a return to the city, lovers hunched together before bed, late-night regulars of various kinds—could never have guessed the role of these nondescript men, or the utter seriousness of their talk.

The widows' meeting turned out to be a watershed in the Trade Center recovery. Burton and Cote were tough guys, accustomed to seeing life as a struggle, and they would not have been unjustified had they responded impatiently to the encounter. This was dangerous to admit out loud, but it was on many people's minds: the firemen's widows were victims of victimization itself, and in their agony and myopia they were starting to blunder around; moreover, they clearly did not represent the thousands of others who had lost family on September 11 and were coming to terms with the events more stoically. It would have been understandable, therefore, if Burton and Cote had mentally stiff-armed the widows, privately dismissing their emotions as overblown and rededicating themselves to the efficiency of the excavation. They had it within their power to do this—and had they been officials in many other parts of the world, they probably would have followed such a hard line. It was lucky for the ultimate success of the recovery effort that this was not the way they naturally reacted.

Instead, over a couple of beers they talked for the first time since September 11 about people's emotional reactions to the attack, and they questioned why they themselves had felt so little affected by the death and destruction at the site. Burton called Cote a "cold fish." Cote pointed out that neither of them had family or close friends who had died. It also had to be admitted that the project was going well, and that for both of them it was utterly consuming professionally, offering an emotional advantage that others did not have: they simply did not have time to dwell on the tragedy. Still, each had been moved that night by the suffering of the widows, and had been troubled by the realization that, though they had tried to do the best possible job, there were people who now believed that their actions were wrong, even wicked. It made them question the doggedness of their approach, and reminded them of a simple imperative that in the crush of daily decisions they were tending to forget: that the unbuilding was more than just a problem of deconstruction, and that for the final measure of success they would have to take emotions into account. They finished their beers, drove downtown, and walked through the site.

In practice the firemen had lost the fight, but the terms of the peace would have to be generous. Burton knew this now, and Giuliani did too. The mayor increased the search teams' numbers back to seventy-five per shift, though they would have to proceed on a less ambitious basis than before: it was understood that beyond being allowed to search for their dead, they would in fact have little say. The tensions never went away, and indeed escalated toward the very end. But already on the morning after the widows' night, at the joint meeting at PS 89, one of the fire chiefs unintentionally made a show of his loss of power when his only contribution was to ask for a moment of silence for the dead of the Queens airline crash—a strangely irrelevant request that emphasized the changes under way by signaling that he had little to say about the work at hand. The widows would be heard from again—but increasingly through formal channels created for them. Mike Burton was now unchallenged as the Trade Center Czar. But he seemed to understand that to succeed he would have to keep his ambitions in check, and that America does not function as a dictatorship of rationalists.

And so the recovery proceeded, not as a united or a heroic exercise but as a set of accommodations worked out among self-centered groups sharing a pragmatic understanding that this was an important job, and that it was primarily a physical one. The only solution was to attack the ruins—to cut and drop the skeletal walls, lift the heavy steel, chew up the rubble mountains, pause to recover the dead, demolish the stump of the Marriott hotel, tear down the burned-through hulks of Buildings Four, Five, and Six, stabilize the neighboring high-rises, shore up the damaged subway, excavate the insanely packed foundation hole, reinforce and protect the slurry wall, and run a fleet of trucks and barges to haul the debris away. About halfway through Ken Holden said to me, "Excavation, remains, recovery, removal—repeat," because in essence that cycle constituted the work. But of course the complications were considerable.

Indeed, the project defied most attempts to impose order, and it eventually defeated even the crudest organizational plans: the four-quadrant overlay, for instance, which slowly faded away; or the initial assumption that the excavation could proceed more or less uniformly from top to bottom, which because of the variability and interconnectedness of the debris it could not; or Mike Burton's related hope that the aboveground work might be finished by Giuliani's last day in office, December 31, 2001, which was far too neat to be even remotely possible. To the very end of the recovery effort and beyond, into the disputes over rebuilding that followed, nothing about the Trade Center site was tidy. Furthermore, as the work proceeded, the site became rougher and more complicated rather than less, a human landscape no longer shaped by its cataclysmic birth so much as by all the initiatives that had followed—histories measured in horsepower and steel, too obscure to remember or write down. To his credit, Mike Burton understood this. When he roamed the pile, as he did twice each day and once again at night, he seemed to accept the disorder there as being in the nature of an energetic response. Rather than hunting out infractions or putting a stop to unauthorized work, as a less confident ruler might have done, he watched for what he called "dead real estate"—unexpectedly quiet ground that resulted from supply-line breakdowns, trucking gridlock, or simple miscommunication between crews that worked the day shift and those that worked the night.

Sometimes he went without a hardhat, and often without a

respirator—flouting the safety rules not just, it seemed, to show his toughness and rank but, more important, and to the exclusion of pettier concerns, to demonstrate his dedication to the real work of the site. This was the attitude he expected from others, and the message got through. Jan Szumanski, a superintendent with Tully Construction, once joked to me that Burton didn't need the protective equipment, because people like him didn't have a heart or lungs. In the context of the Trade Center site, Szumanski meant it as a compliment.

For a man with a comfortable house in Chappaqua, Burton was surprisingly at home on the pile. He could pitch his voice perfectly to make himself heard over the cacophony there—the roar of a hundred diesels, the hiss of cutting torches, the screech, rattle, and bang of extracted steel. He was too focused on the job ever quite to seem relaxed, but he was less skittish than many others, and did not hesitate to wade through the billows of smoke and dust, or to climb the steep slopes of loose debris. He would stand in the thick of the action, sometimes for an hour or more, with his arms folded and a frown of concentration on his face, taking it all in. His presence was crucial, especially in the fall and early winter, when the Trade Center crews were operating largely blind, guided only by common sense and instinct, and encountering dangers that had never been seen before. It was natural if occasionally they equivocated, or shied away from problems. Reliably, however, Burton never did—and instead had the habit of going out and actively looking for trouble. Most of the problems he discovered were so specific to the Trade Center site—and in that sense technical—that they would have been invisible to outside observers. Burton would have been invisible to those observers too: an unexceptional-looking man, moving among the more dramatic crowds, often accompanied by a trusted adviser (Bill Cote, for instance, or Peter Rinaldi), but generally making little effort to display his power. The work that he did on the pile was impossible to photograph, and difficult even to quantify, because it amounted largely to the avoidance of negatives. Nonetheless, it's certain that without Burton's frequent excursions there, those negatives would have added up.

Late one night in the fall we stood together on a debris hill, looking down across a scene that was at once familiar and sur-real—the smoke and steam rising under harsh stadium lights, the orange sparks cascading from cutting torches, the teams of firemen standing in clusters with their rakes and shovels, the diesel excavators restlessly in motion, doing their dinosaur danc-ing in boiling dust and fields of flame. I said, "It's your kingdom, Mike. It's your empire." He laughed, and didn't hide his satisfac-tion. He seemed to be savoring the moment. Nothing had gone really wrong for a while, at least not to the extent of threatening Burton with failure. The dead were at last known to be dead. The protective slurry wall had been successfully stabilized. And the process of excavation was so well under control that the main thrust now was a pile-improvement project, Burton's initiative, to build a temporary access road in the Bovis quadrant, from West Street into a deep depression, known as the "valley," that lay at the center of the ruins between the footprints of the North and South Towers. The road was being built of ground-asphalt millings laid on top of the debris along a route that sloped steeply downward from the street. Over in the southeast corner, in Tully territory, Jan Szumanski was coming on aggressively with a sec-ond access road that, after S-turning around and through the steel remnants of the South Tower, would eventually link up with the Bovis road. The imminent arrival of debris-removal trucks at the heart of the ruins signaled a crucial shift in the geometry of the operation. The trucks would be loaded at the center of the pile—an important improvement over the current reliance on long daisy chains, in which the grapplers and excavators passed each piece of steel down the line and into the waiting trucks on the street.

Henceforth the excavation would proceed with increasing efficiency from the center outward, following a pattern similar to that of open-pit mining—albeit complicated here by variations in the ruins, by the need to preserve the slurry wall, and, of course, by the presence of the human dead. Those dead constituted an argument against excessive road building, because of the certainty that some of them were being buried under the millings, that their discovery would be delayed, and that for months to come, while their remains continued to decay, heavy trucks would roll across their graves. Burton had learned his lesson from the firemen's widows, and he was determined to keep the emotional factors in mind. Nonetheless, when he looked ahead to what would soon be a deep hole in the Manhattan ground (increasingly defined by the sheer faces of the slurry wall), he saw no alternative to the roads for the removal of the debris. It was understandable, therefore, if that night when I stood beside him he allowed himself a moment of satisfaction with the progress being made.

But he was not a man to luxuriate for long, and with much of the debris-removal operation awaiting the completion of the Bovis road, he remained sensitive to signs of trouble. For instance, he turned now to a problem he had already encountered in the afternoon by the ruins of the Marriott hotel, where three diesel grapplers had been picking at the debris in a loose daisy chain, without contributing to the more important work in the valley. Burton had spoken about it to a Bovis field supervisor, and the grapplers had joined the main effort, but word had clearly not been passed to the night shift, because the machines were once again squatting in the corner on a mound near the hotel. To make matters worse, two of the three had just broken down. One of them was a giant "750" with a long-arm attachment that had been pulling at the top of the ruins when a steel beam had snapped, cutting one of its hydraulic lines. Bleeding badly, the 750 began an awkward retreat, until, having lost 120 gallons of its vital fluids, it died. Burton was extremely annoyed. He clambered through the ruins until he found another supervisor, and with a smile that left no doubt about his anger, he calmly suggested that the company pay better attention.

But stronger words were often required—work for which Burton was not well suited, and which he gladly left to the site's principal "pusher," Lou Mendes, the DDC's assistant commissioner for special projects. He was a blunt bulldog of a man, of Portuguese origins, whose technique for keeping things moving was to stand toe-to-toe with the other big guys on the pile and systematically lose his temper. Mendes in action was a phenomenon to behold. He had a large, jowly face, a low forehead, and small eyes, which gave him a glowering look even when he was feeling relaxed. When he was angry, his face darkened and became at times mocking, facetious, outraged, pained, or openly disbelieving. He would take on self-respecting men and ridicule and scold them as if they were delinquent children. At the most extreme, during a middle period when AMEC's operation was stumbling, he grew so aggressive that he seemed to risk a complete breakdown in communications, if not outright physical assault. At PS 89 one day, after another slow night in the AMEC quadrant, Mendes called in the various vice-presidents and managers from the four prime contractors, sat them down in an abandoned classroom, and proceeded for nearly an hour to humiliate AMEC about its lack of performance. He said, "This is going to be a twenty-four-hour operation with or without AMEC. I go around, I look at Tully, I look at Bovis, and they're productive all night. I want the same out of you." The problem, as usual, seemed to result from a lack of communication between the shifts. "I want to know what's the matter with you guys. Don't you talk to each other at all? You keep this up, and I'll shut down your whole operation. You watch me. I'll throw you off the job."

The AMEC people tried to explain: certain equipment had broken down, a supplier had delivered late, there were unexpected complications with the steel, the Fire Department had not shown up with a fire-suppression team for the scheduled cutting of the steel spears from the wounded American Express building. Mendes would have none of it, and repeated the excuses loudly and incredulously, as if to allow others in the room to savor the incompetence on display here. He said, "If you need equipment, we'll lend it to you. You need a backhoe? Between Tully and Bovis, I'll get you what you need." He seemed to be implying that AMEC would have had trouble acquiring even a lawn tractor. The AMEC people listened with long-suffering expressions. They were of course seething at the treatment (and they later complained, to no avail), but they knew that any attempt to expel them from the job, even if unsuccessful, would seriously damage the reputation of the firm—an unexpected wrinkle in the otherwise proud claim of involvement in the World Trade Center recovery. Mendes in that sense had extraordinary power, and they had little choice but to accept his verbal lashings. He said, "I don't care if your night guys are late or not. I don't care about your dinner plans. You stay on and talk to the night shift before you leave. You make sure they understand the plan. And then talk to me. You *call* me before you leave." He pointed at himself. "Lou Mendes," he said, and rattled off his cell-phone number. "Did you write that down?"

"Yes, Lou."

"And I want you to bring in lights. Tonight. You have a twenty-four-hour operation. I want to see lights. I want to *see* a twenty-four-hour operation. Assholes and elbows. Do I make myself clear?"

"Yes, Lou."

Burton and Holden stayed away from such messy scenes, but each of them quietly insisted to me that Mendes was a good and necessary man. It turned out that he was a sentimentalist, too (of course), and not quite the bully that at first he had appeared to be. Even those on the receiving end of his attacks, which were by no means limited to AMEC, eventually came to recognize the legitimacy of his complaints, if never quite his methods of delivery. After a while a sort of fondness characterized the exchanges.

AMEC solved its problems and began to make rapid progress on the pile, but then at the center of the action—the former valley that by midwinter had been deepened almost to bedrock and was known as "the hole"—Bovis started having trouble. The problem this time was with restricted maneuvering space in the loading zone, as a result of which near gridlock had set in, and the output of steel and debris (known as "production") had dropped from a high of 12,000 tons a day to merely 2,000. On the final inspection line at Fresh Kills, where hundreds of law-enforcement officers were being retained to handle a larger flow, the people in charge were complaining about the lack of predictability, and demanding to know if they should plan on such low numbers in the future.

Mendes was outraged at Bovis. He called for a meeting with a dozen men he considered to be the key players on the pile—Peter Rinaldi, Sam Melisi, Pablo Lopez, and managers from the most important contractors. Once they were all gathered, sitting around a small table or perched on the windowsills, Mendes said, "Shut the door." Dispensing with the niceties, he immediately went on the attack. He called the work in the hole a "circus." The field superintendent from Bovis, a ruddy man named Charlie Vitchers, who had a reputation for practicality and competence, tried to explain that the problem was due partly to the lack of clear command: someone needed to draw up a chart to show how authority was supposed to flow. Mendes grew apoplectic. "A chart? A goddamned chart? We've got charts coming out the ass. We've got charts of charts! We've got charts to *make* charts!"

"Maybe we should . . . "

"It's all maybes! I don't want maybes. I want to know what you guys think you're *doing* out there! What, digging holes and filling them in? Building Six is on hold! The donut is on hold! Greenwich is on hold! The whole goddamned *hole's* on hold! My job is to get this job done! You may not like me. You may think I'm a pain in the ass. But my job is to do the job! I want to know what the hell you think your job is!"

Charlie Vitchers's face was getting ruddier. He worked visibly to maintain his calm. He said, "We got ourselves into a cluster fuck. We knew it. We saw it coming. We knew we'd be in it for a week. But it's getting better now."

One of the Tully heirs, Jim Tully, was in the room. He was a lean, quiet man with hooded eyes and a pockmarked face, the very image of a New York street tough. Of the siblings who were running the company, with the hands-on approach for which the family was known, he was the field man, the one most at home on the pile. Currently he was managing the trucking and heavy-equipment operation as a subcontractor under Bovis, in one of the numerous shifting arrangements that had blurred the original quadrants. Incongruously, he now tried to placate Mendes. He folded his hands on the table as if to show that he was unarmed. He said, "Look. Lou. I'm a production maniac too. I mean, you're not the only guy, Lou. It's been burning holes in my gut all week. So maybe you can understand—just a smidgen?"

Mendes was in top form. He said, "Maybe I understand 'just a smidgen,' I'll give you that. But this whole operation is *bullshit*!"

Sam Melisi was sitting at the far end of the table observing the argument with his fellow firemen in mind, as always, and by his mere presence reminding the people there of values other than speed and ambition—particularly the importance of allowing sufficient time to find the dead. Melisi had questioned Mendes's methods at least once, in his apologetic, unassuming way, voicing to Burton the firemen's complaints that Mendes was running roughshod over the inspection process in the hole. Mendes had immediately backed down; there was something about Melisi's moral authority that even Mendes could not resist. Jim Tully seemed to want to borrow some of that now. He said, "Plus, yesterday was a very tough day for the firemen. There were a lot of recoveries."

Melisi said nothing. He would have called it a good day.

Mendes refused to get dragged into the conversation. He pointed his finger at Tully and bellowed, "Jim! I want guys pushing trucks! Today! Tonight! Trucks! You got it?"

"Yes, Lou."

"I'm only a city fucking worker. You guys are the experts! So what are you doing about it?" He glared around the room, and repeated himself more quietly: "I'm just a damned civil servant."

Then he paused and said, "Well, maybe I'm not so civil." People laughed, though it was more a verbal reflex than a joke.

Charlie Vitchers said, "Okay, Lou. Tonight I'll read them the riot act. We'll see what happens."

Mendes was spent. He said, "Push them, Charlie. They're not made out of wax."

But the problems of the project generally had more to do with too much motion than with too little: confusion, lack of focus, gridlock, risk-taking, the chaotic, free-for-all nature of the emergency response that was also its genius and strength. After years of struggling with the New York unions, some of the construction managers had a hard time understanding that inside the inner world people's attitudes had changed, and that the same workers who couldn't be made to care beyond their paychecks about erecting the next shopping mall or office building cared inordinately about pulling these ruins apart. Not that they worked here for free—or could have afforded to. After the site matured, the only volunteers in that sense were some of the Salvation Army and Red Cross people who served food in a large white tent known as the Taj Mahal, a communal space that, by offering the workers, firemen, and police officers a place to sit down for free meals, did as much to soothe wounds and maintain the peace as any political bargaining or formal compromise. Those volunteers were restricted to the outer zone of the site—the staging area where the debris trucks queued up and, when loaded, were tarped and driven through the washdown stations on their way to the piers, and where a large but conventional effort was under way to clean and reconstruct the variously damaged buildings all around. The pile itself was open only to the elite inner-core crews, and it was strictly union ground—but not as New Yorkers had seen union ground before. The mob shied away from the job. There was little featherbedding, and no strong-arming of the New York kind. With the exception of the firefighters' action, which was a special case, there were no threatened walkouts or strikes.

Because of the physical unknowns of the debris, as well as the frequent interruptions for the recovery of the dead, the work was done on the basis of open-ended "time and materials" agreements, as opposed to the standard packaged bids, and though some of the truckers cheated, and certain contractors grossly inflated their costs, on the whole, workers never got the idea to slow down and take advantage of the federal largesse. They were by no means perfect. There was the looting, of course. And Burton threw an entire AMEC crew of twenty-seven ironworkers off the site for spending a shift doing nothing at all. But Peter Rinaldi, who for all his adult life had overseen construction projects for the Port Authority, regularly expressed his amazement to me at the efficiency of the operation. Never before had he seen a time-and-materials job that functioned so well. The overtime helped: the heavy-equipment operators, for example, were earning at a rate of up to $200,000 a year, and some of the firemen and the police, whose pensions were based on their last level of pay, were experiencing such a windfall that financial logic required them to retire after leaving the site. But it would be a mistake to think that people were going around silently counting their blessings. They were in fact responding to the attack with all their might. They were doing the country's dirty work, and somewhere in the background the money was flowing into the bank.

The stars of the show were the machines themselves, and particularly the big diesel excavators, marvels of hydraulics and steel, which roamed through the smoke and debris on caterpillar tracks and in the hands of their operators became living things, the insatiable king dinosaurs in a world of ruin. They came in various sizes, from the "small" 320s (which could pull apart an ordinary house in minutes) to the oversized 1200s, monstrous mining machines rarely seen in New York, which proved to be too awkward for many uses on the pile. Most of the work was done by the 750s—sufficiently big, sufficiently lean, enormously persistent beasts that battled the debris without rest. Each 750 weighed in at 180,000 pounds (as compared with 140,000 pounds for the heaviest trucks, fully loaded), and was equipped with an articulating arm and one of three hydraulically powered attachments—steel-cutting "shears" (often attached to an extra-long arm, for reaching high or wreaking havoc deep inside the standing ruins); conventional "buckets," useful at the lower levels of the pile in areas of pulverized debris; or, most often, "grapplers," gap-toothed claws that could open eight feet wide, but could also close into an overbite so tight that it could snap twigs. The grappler-equipped excavators (known simply as grapplers themselves) dominated the battle until it moved well below ground. Working fast and in tandem, the machines picked the ruins apart one piece at a time. The steel they took on included the heaviest ever used in a building—box columns weighing 3,100 pounds per foot, so that a merely man-size length would amount to almost 19,000 pounds, and a fifty-foot section would come in at nearly the weight of the grappler itself. Some of the loose steel could be "flown" out by the enormous cranes that ringed the pile—but the process was so tedious that it was reserved for special cases: for instance, for lifting beams during the search for survivors among ruins too rough to allow the grapplers access, or for the dismantling of the skeletal walls, most of which were torched apart one section at a time by ironworkers in suspended baskets, and then lowered gently to the ground. This left the bulk of the fight to the grapplers, which were just tough enough to take it on.

Actually, of course, it was the operators who accepted the fight. They were said to be the best in the business, and this was easy to believe. At the start of a shift they didn't just climb aboard and sit down but seemed, rather, to strap on the equipment much as good pilots strap on their wings. Like those pilots, too, they were artists of motion—fluid, expressive, and intuitively at one with their machines. The cabs that they sat in were enclosed in wire mesh and sound-dampening, shatterproof glass; they had single comfortable seats, filtered and cooled or heated air, automotive stereos, and a combination of pedals and tightly coupled, variable-rate joysticks that allowed the operators nearly bionic control. It was the control especially that gave the grapplers their beauty. The operators might drive to work like ordinary commuters, frustrated by traffic, by parking regulations, by lines at Starbucks for insipid coffee; but after they settled into their machines, they could put all that aside, and go rumbling off into the faraway land of ruin; and if they came to terrain too wild to cross, often they could build a way through; and when they came to the field of battle, typically among other grapplers straining there, they could reach their own arms out twenty feet, clamp their own steel claws around multi-ton splinters, and with fire and smoke erupting, while shuddering and rocking forward onto the toes of their tracks, they could wrestle those splinters clear. They could also then stretch their claws wide and angle them up to use just the bottom fingers to gently stir loosened debris for the firemen to see. The operators had all that power and grace at their command, and they possessed more imagination than ordinary construction jobs had let them exercise before. Now they had been given a high purpose, and been told roughly what Sam Melisi had been told: just go and see what you can do. It was a liberation, because they knew they could do a lot. They were resourceful. They were like pioneers.

Taking risks was a necessary part of the deal. That was true for others there too: even when they were not breathing in the smoke and dust, or climbing across treacherous slopes, the workers worked on top of weakened, partially collapsed structures, which bounced and shook underfoot, and sometimes gave way. But the grappler operators were particularly exposed, not only because they led the effort, often balanced precariously and pulling on unstable debris, but also because they seemed to feel protected by the mass of their machines. This was an error. The grapplers were like dinosaurs, but with thin skins. On two occasions when they ventured into areas from which Peter Rinaldi had excluded them, the pile suddenly collapsed, dropping them into voids. The two machines were badly damaged. It was a matter purely of luck that neither of the operators was injured. Rinaldi told me he would be very surprised if no one got killed at the site. On another occasion I was with him when he noticed to his alarm that one of Tully's 750s had nosed into a corner of the partially collapsed Building Five, and was reaching up and pulling heavy debris down from overhead. Even if the building itself did not let loose, the pieces that were falling could have sliced right through the cab. Rinaldi got on the radio, which didn't work. We walked across the pile to find Jan Szumanski. After a period of anxiety and confusion the grappler retreated.

With us at the time was an affable, boyish-looking man named David Griffin, age thirty-four, who was the project's chief demolition consultant and its token Southerner. Demolition in this context meant the unbuilding not of all the ruins but of the standing structures, of which there were plenty—Buildings Four, Five, and Six; the stump of the Marriott hotel; and all of the intact and partially intact basement levels of the foundation hole. Griffin, as usual, was wearing immaculate golfing clothes. His site ID tag dangled from a neck strap embroidered with "I Love Jesus." He was less concerned than Rinaldi about the errant grappler, having seen much worse in his professional life, but he had gentlemanly manners, and said nothing about that now. Instead he reflected on the Trade Center work in general. He said, "It's like playing Jenga—who's gonna pick out the last piece? Only thing is, instead of falling down on your kitchen table, the pile falls five levels down—and you're in it." I asked Griffin if he was good at playing Jenga. He grinned and said, "Yeah, I reckon I am." It was typical Griffin. He was a personable guy. He came from faraway Greensboro, North Carolina, and was the butt of a lot of NASCAR jokes. Indeed, his uncle was a NASCAR official. Griffin thought it was a pretty big deal. But he was not a complete rube. He knew all about the stereotypes, even found them funny. Once at a morning meeting at PS 89, when Griffin asked about the value of the bullion that was soon to be removed from the vault under the ruins, Burton said, "It's enough to buy all of North Carolina—and they'll throw in South Carolina for free." Griffin laughed along with the rest.

But the bullion was worth only about a quarter billion dollars, and Griffin's own daddy, as he called him, was probably worth that much or more. His daddy's name was David too, though people called him "D.H." He was a tobacco farmer's boy, a ninth-grade dropout who in 1958, when he was nineteen, went to work on a cigarette production line, where he met his wife, and where he thought he'd spend his life. In his spare time he sold parts off a dozen junked cars in his front yard. In 1959, he tore down an old church to salvage the lumber and nails, and did it so well that the city inspector (who stopped by because D.H. lacked a demolition permit) asked him to handle a small apartment building next. D.H. soon realized that the value of demolition lay as much in the resale of salvaged materials as in the service rendered, and, as he liked to explain, he just kept going from there. By the time of the Trade Center collapse he had become one of the three largest demolition contractors in the United States, one of the largest scrap-metal dealers in the South, a construction magnate, and the master of more businesses and real estate in North Carolina than even he could remember.

New York was not his kind of town, and he had opposed David's impulse to get involved, but after the Griffin name became attached to the Trade Center project, D.H. arrived twice at the site to see the work, of which, of course, he was proud. He was a burly, gray-haired man of sixty-two, with a bad knee, a strong Carolina accent, and unusually quick eyes. David took him around on an all-terrain "Gator." D.H. was friendly, but didn't have much to say to the New Yorkers who passed by. He wanted to talk to David about family, and a little business. He told me about his weekend relaxation, which he called "riding around," and which amounted to an activity rather close to his work—a Sunday-drive form of looking for trades, in which the size of the deal didn't matter so much as its quality. David laughed and said that just last week his daddy, having come into 20,000 T-shirts (D.H.: "Class One T-shirts. Orange ones, white ones, blue ones") at 25 cents apiece, had taken a few boxes of them to the parking lots of local shopping centers, and for a dollar each had sold dozens out of the trunk of his car. His customers must have figured he was an old man down on his luck. Even D.H. could now chuckle about it. But later that afternoon, as we talked, I got the distinct impression that he was feeling out my interest in acquiring, say, an antique shotgun, or maybe a knife.

When the Twin Towers collapsed, D.H. Griffin, like most Americans, was so wrapped up in his affairs that he was unable to respond beyond the standard feelings of anger and surprise. David Griffin was a rich and busy man in his own right, president of the demolition company, with a wife and young children, but not quite so established as his father. He was drawn to the Trade Center site much as Ken Holden and Mike Burton were—because he had knowledge to offer, and could not stay away. He had a business justification, too: on Thursday, September 13, he called his father and said, "Daddy, I'm going to New York. Because this is the kind of job, if it breaks—when it *does* break—they're not gonna pick up the phone book and look under 'Demolition Contractors' to see who could help them. I'm gonna leave, and if I don't have anything by Monday, all I'll feel is I wasted a weekend in New York. Nothing ventured, nothing gained."

His wife, who had been to New York only once before (for a day), insisted on accompanying him, and on bringing their children. They loaded up their Suburban, drove nine hours north, and in the middle of the night checked into the Palace Hotel, in midtown Manhattan. Griffin got a few hours' sleep, and early Friday morning, in a pouring rain, he drove downtown as far as he could, grabbed a raincoat and a hardhat out of the back of his vehicle, hung a respirator around his neck, and started walking. When he came to a police barricade at Canal Street, he did not hesitate, and he passed through unchallenged. His heart pounded. He thought, *I'm in! I'm in!* But four blocks from the site he was stopped at another barricade—this one manned by the New York National Guard. A soldier said, "Where do you think you're going?"

"I'm going to work."

"Well, lemme see your pass."

"I don't have a pass."

"If you don't have a pass, you're not going in. Who are you working for?"

"Well . . . Bovis." Griffin had contacts there, some names to look up.

The soldier said, "What, you don't have a pass or anything?"

Griffin said, "I come from North Carolina."

That fixed it. The soldier said, "No pass—you can't go in."

Griffin stood around, unsure of what to do. The rain came down. People with passes went through. After a while a group of Red Cross volunteers showed up with drinks for the soldiers. The soldiers began talking to them. Griffin drifted inside the barricade. The soldiers didn't seem to notice. He walked toward the pile, and the next thing he knew, six months had gone by.

His Bovis contacts knew him by reputation: this was D.H.'s boy, a kid who'd grown up at wrecking sites from the age of two, sleeping in concrete culverts at night, with laborers standing guard. The Bovis people understood the value of the homegrown experience that Griffin could provide. They hired him for their quadrant, and introduced him around. For the first several weeks he had competition from a more flamboyant demolition man, who ultimately flamed out in the eyes of the site managers at PS 89 and was asked by Ken Holden to leave the job. People had trust in Griffin by then, as much for his lack of grandstanding as for his obvious technical competence. At Burton's request he took on the role of demolition consultant for the whole site, planning and managing the dismantling of all the standing structures. It was a crucial position, central to the entire project.

Toward the end, over a sandwich with his father and me, Griffin described his early thoughts. He had considered some selective blasting at first, to drop the skeletal walls and bring down the ruined buildings. He called it "shooting" them. Holden and Burton vetoed the idea, primarily because of the jitteriness of the neighbors. As much to his father as to me, Griffin said, "These New Yorkers just flip out. You know, this ain't South Pittsburgh."

D.H. said, "Nope."

Griffin figured he would have the space to use a wrecking ball on Buildings Four and Five, and indeed that is what he did, after pre-cutting the internal structure of the buildings, a dangerous process he called "juicing them up." It was an unusual procedure for New York, but faster and cheaper than the city's standard labor-intensive technique of incremental deconstruction.

He also figured correctly that he'd be able to topple the north skeletal wall, a towering structure that had broken at the base and was leaning heavily against Building Six. This was done with five grapplers arranged side by side, pulling simultaneously on cables bolted to the top. Eventually he was able to rock and topple Building Six, too. But for the most part, such wholesale wrecking techniques were not going to work at the Trade Center site. Griffin said, "I knew there wasn't but one way to do it."

D.H. said, "Yep."

"That's to start cutting and pulling it down."

He meant piece by piece, and it was no big deal. He had a way of making things seem simple. He brought in a few of his own people—Southern boys all, who had the hardscrabble background necessary to get the job done. They moved into some vacant apartments nearby. They were the most undaunted workers at the site. Months later Burton expressed surprise over beer one night when he discovered that Griffin had only a high school education. No one had bothered to ask before, and afterward no one thought it mattered.

The truth is, what Griffin had grown up knowing about risk and safety, others at the site, more formally educated, were now having to learn. Regulation was simply not possible at the start, and even after it began to creep in, its real purpose was to exist

officially on the books, playing a rearguard position while the
project surged ahead and continued to allow personal responsibility and individual choice to prevail. Peter Rinaldi adapted to
the freedom naturally, despite his years spent within the confines
of the Port Authority, and he became Griffin's ally in the ordinar-
ily cautious camp of the engineers. Others had a harder time accepting risk, but even they eventually came around. There was,
for instance, an encounter that became well-known at PS 89, between Griffin and a DDC engineer who was nominally responsible for the aboveground structures, but whose penchant for
memo-writing made him something of a misfit at the site. For
weeks Griffin had been pushing the engineer to let him proceed
with the demolition of the Marriott ruins, which rose three stories above the street and extended all six levels down into the
foundation hole. There was no safe way to take those ruins on.
Access from the sides was impossible because of the proximity of
unstable structures and of the vulnerable slurry wall. Griffin's solution—to put an excavator directly on top and wreck the Marriott from above—was obviously very risky. The DDC engineer
kept blocking the action. At last Griffin came as close to losing
his temper as he ever did. He said, "We're not going to *talk* it
down. We gotta do *something*."

The engineer finally gave in. He said, "Okay. I wouldn't *rec-
ommend* it, but I guess it's okay." That formulation quickly went
around. It described how the site actually worked—by people
turning a blind eye.

Given the go-ahead, a grappler operator immediately drove
his machine to the top of the hotel and, at conscious risk to himself, began to tear it apart nearly underfoot. It was dangerous behavior, but typical on the pile. Griffin himself, who often stood in
the thick of the action, once had a narrow escape when the coun-

terweight of a swinging excavator hit him hard in the back, luckily knocking him down a slope rather than against the debris, which would probably have been fatal. He checked with the site doctor, and went back to work. At another time, when a large section of steel unexpectedly fell, a fire chief came rushing up to Griffin yelling, "Where's the safety zone on this job?" and Griffin calmly responded by naming the site's outer-perimeter line: "Chambers Street. Do you want to close the whole place down?" The fire chief got the message. Most people eventually did. Risk was the very nature of the Trade Center operation.

This was sometimes difficult to grasp for people on the outside, where there were flurries of news reports about worrisome safety violations at the site. Outsiders believed that the constant danger—along with the presence of the dead—had to be getting to people. One afternoon when Pablo Lopez stopped by the midtown engineering offices of Thornton-Tomasetti, he was offered a session with a consulting psychologist, a woman whom the company had dutifully retained to assist its Trade Center crew. The psychologist asked Lopez to make himself comfortable. She had a slow, soothing way of talking, which had the immediate effect of irritating him. He told me about it the next day, exaggerating the intervals between her words.

He said, "She says, 'Close . . . your . . . eyes.'

"So I close my eyes. Okay, now what?

"She says, 'Imagine . . . a . . . safe . . . place.'

"I think, safe place? What's that? At least she could have said 'Imagine a steak house.' I mean, where've you been, lady? I live in New York City, and there's an anthrax scare going on! I go home, and my wife is ironing the mail! And where is it I work? It's underground in the World Trade Center."

I don't know what Lopez said to the psychologist, but his

point to me was that he wasn't planning to move away. He lived at the center of the world because he liked the action. He worked at the Trade Center because he wanted to. He wasn't searching for safety. He didn't need to close his eyes, or to make himself comfortable. He didn't need the teddy bears that volunteers kept handing out. And he wasn't afraid of the dead.

In the end, 1.5 million tons of ruins were extracted from the seventeen acres of the Trade Center site. The vast bulk of the material was barged twenty-six miles to the Fresh Kills landfill for sorting, final inspection, and burial. Despite the negative emotions it evoked in Manhattan, Fresh Kills was an excellent choice for the work—one of the largest open spaces in New York City, a magnificently barren landscape of earth-capped refuse, spreading across 2,200 acres and rising in places about 200 feet above the tidal estuaries of Staten Island. It had been retired from service six months before, in March of 2001, presumably to become a park someday, but had been reactivated for this one final purpose. Now again it was a dump, and one of the largest in the world. But it offered complete privacy and calm, and allowed for surprising dignity during the sad and gritty operation to come.

For the landfill itself, the prospect of accepting the ruins did not pose a significant challenge. The docks and equipment necessary for offloading the barges remained in place, as did even the little service boat that opened and closed the floating boom that kept flotsam from drifting away. Moreover, no matter how rapidly the Trade Center excavation proceeded, on the scale of New York City garbage its output was not expected to be unusual; in fact, the disaster site's highest level of daily production, at 13,900 tons, was only a thousand tons more than just the household rubbish that Fresh Kills had dealt with in its last years—a level that itself was merely a third of the city's solid-waste production, estimated at 14 million tons a year. In strictly material terms, therefore, the Trade Center debris was hardly more than just another curiosity—a variation in the urban detritus that Fresh Kills had accommodated for a half century. But of course in emotional terms the Trade Center was very different stuff.

On 176 acres across the top of its highest hill, Fresh Kills set aside an area for the operation. The ruins began arriving by truck the very first night. The initial loads were of high-grade structural steel—torn perimeter sections of the Twin Towers, and fractured and twisted beams from the inner cores. At the site they had been inspected for the dead, but a few nonetheless contained the occasional human remains. These pieces of steel arrived in a rush that echoed the urgency of the search for the living, and they accumulated uncontrollably in piles so heavy that they began to crush the hill, damaging the system of subsurface pipes that channeled methane gas from the site. Fresh Kills cried for relief from the weight and soon got it, when for independent reasons an agreement was worked out to sell and send the structural steel from the Trade Center site directly to scrapyards in New Jersey. As much to manage the quantities as to turn a profit, those scrapyards wasted no time in cutting up the steel for further resale and shipment. Fresh Kills was able to unburden itself by shipping its heavy steel to those same yards, but the accumulation was so great (and the pieces so dangerous and difficult to handle) that the process took months to complete.

Meanwhile, the barges had been put in motion, the Trade Center debris was arriving by the thousands of tons daily, and production-line procedures for handling it were proving up to the task. Those procedures remained remarkably stable over the course of the operation, because though the conditions at the Trade Center site frequently changed (resulting in production spikes), once the heavy steel had been redirected to New Jersey, the nature of the output destined for Fresh Kills remained largely the same: bargeloads of rubble consisting of broken and crushed concrete, asbestos, asphalt millings, rebar, and other forms of light steel—all stirred through with a homogenized mixture of details from 50,000 working lives, nearly 3,000 of which had just ended violently. Fresh Kills' job was to separate the human mixture from the rest—to dehomogenize the debris.

The process started intuitively on the barges themselves, where some of the tugboat crews believed they could judge the organic content of the loads from the seagulls overhead, scavengers who were drawn by odor but had little chance to feed, and whose flocks diminished over time. Throughout the fall and into the winter some of the debris was so hot that, fanned by harbor breezes, it smoked and burst into flame. When it arrived at Fresh Kills, it was lifted by giant excavators into specialized dump trucks, which drove it up a curving dirt road to the top of the hill and released it into little mounds, where the sorting began. The hilltop was a wild-looking place, with American flags whipping in cold winds, like the outpost of a government expedition to a toxic planet. It was scattered about with heavy equipment, truck trailers, and prefabricated structures of various kinds, and roamed by hundreds of workers (typically police officers or FBI agents) who were garbed in white protective suits, respirators, gloves, and high rubber boots. For visitors first arriving from the Trade Center site, where people worked largely unprotected, the clothing in particular seemed odd, as if something must have happened to the debris to make it more dangerous on the way over. Otherwise there was plenty of evidence that workers in both places were handling the same materials: all around stood huge piles of Trade Center debris—much of it now sorted, inspected, and awaiting burial—that elicited unexpected feelings of familiarity and later even of fondness, like old acquaintances encountered in a foreign land. The hilltop was of course a part of America, and by geographic measures it was not far removed from Manhattan: on a clear day from there you could even count the monuments of the skyline, minus two. But it was isolated and exotic nonetheless.

Certainly no one had ever run a dump this way before. After the debris arrived from the piers, the large metal pieces were extracted, most significantly the vicious tangles of rebar "spaghetti," which by hanging untrimmed off trucks had threatened workers at the Trade Center site with decapitation and now in due justice were to be sliced, sold, and melted down. Once the metal was extracted, the rest of the debris was scooped up and poured into giant shakers, of which there were as many as four. Debris that was larger than six inches across was removed, spread over a field, and raked through by hand. The remaining materials were fed into equally giant mechanical sifters, which shook and spun the loads into three separate debris streams according to size. The first stream contained material less than a quarter inch across, and (with the exception of the occasional fingernail, as an FBI agent mentioned to me) it consisted almost entirely of asphalt millings and dirt, and was discarded without being inspected. The second and third streams contained the larger debris. These materials were carefully scrutinized. They were fed onto variable-speed conveyor belts that ran through plastic-walled structures where white-clad workers sat on stools along what amounted to disassembly lines, watching ninety minutes at a time for anything that might be assigned to a victim—badges, guns, and Palm Pilots, for instance—or that might be material evidence bearing on the acts that brought the buildings down.

Peter Rinaldi's long-expired Port Authority identity card was found there, on one of the lines: he had discarded it in his desk on the seventy-second floor, and it was returned to him ceremoniously by two Port Authority policemen, who called him with the news and then escorted the card to his office. Most of the personal effects that were found at Fresh Kills were equally unimportant, for the happy reason that 95 percent of the people who had worked in the buildings were still alive. Nonetheless, the searching had to be done. The cockpit voice recorders from the hijacked airplanes were never found. But the inspection process did turn up an average of five body parts a day, most of them very small—which by the time Fresh Kills closed again, in the summer of 2002, had led to the positive identification of an additional seventy-eight missing people. Equally helpful in the emotional climate of the time, the process ensured that none of those particular body parts (and obviously very few others) had been treated disrespectfully or "thrown out at the dump."

The other effect of the inspections, of course, was to free the debris for burial. The interments began right away, in a patch-work pattern across the hilltop, and consisted not of digging graves but of spreading the Trade Center remains and covering them over with a thick blanket of earth. In that most unexpected way the hilltop slowly grew, with the World Trade Center adding rolls and variations to the ground where someday people would come to relax. In fact nothing was just "thrown out at the dump"—not a single piece of those buildings. The work of sorting through the debris was enormous. But it was accomplished, and then at last the job was all over.

On Thursday, May 30, 2002, the closing ceremony was held at

the Trade Center site, and was televised to the world. A 460-foot inclined steel bridge had been built from Liberty Street to the bottom of the foundation hole, which, with the exception of the intact basement structures along the north side, now appeared to be barren and almost clean. The occasion of the closing ceremony was the removal of the last load, a steel column from the South Tower. All the players were there, as usual in separate groups and distinguishable by their clothes: the firemen, the New York City and Port Authority cops, the construction workers, the families of the dead, and, of course, the politicians. Giuliani's Republican successor, Mayor Michael Bloomberg, who was five months into his term, had the sense to keep the ceremony short. The idea, it seemed, was to get this over with, and let people proceed with their lives. There were no speeches of any kind. Among the workers who had been there from the start the mood was a little sweet, like that of a graduation night, when people know but don't admit that they may not see one another again. The steel column lay on a flatbed truck at the bottom of the hole. Peter Rinaldi was down there with it, in a line of others who had been selected for the honor guard. Over the previous months he had continued to gain in reputation and influence, and now, as control of the ground was being returned by the city to the Port Authority, and the subway and PATH commuter-rail restorations were getting under way, he had been asked to stay on and manage the site. It was a strange turn for someone who had built his life inside the Twin Towers, to measure his success by the cavity they had left behind. But he was brimming with self-confidence these days.

Bagpipes played. The truck rolled at the speed of a slow walk. Rinaldi and the rest of the honor guard accompanied the last load as it was carried up the bridge and north along West Street. And that was it, the end—or almost. Actually, several additional

weeks of cleanup lay ahead, during which debris continued to arrive at the landfill and to go through the process of sorting
and inspection. The final disposal proceeded at the normal pace.
The steel was absent. But otherwise, by midsummer, less than
a year after the attack, the World Trade Center and its burned
and pulverized contents lay under bare earth, absorbed, like so
much else of New York's past, into the man-made hills of Fresh
Kills.

But for the residents of the inner world the end had essentially
come months sooner, by early March, merely a half-year after
the collapse. It was a time when few problems remained to be
solved, and the future of the unbuilding had become entirely
clear. The bridge had just been completed, and soon the excava-
tions would start on the last major area of compacted debris, under the Tully access road. David Griffin had packed up and gone
home to North Carolina. Safety restrictions were increasing by
the day. Ken Holden was philosophical about it, and, as his father
might have years before, he played a little word game—something like metaphor-cramming. He said, "When the smoke
clears, the nitpickers come out of the closet." And it was true: the
regulators and auditors had arrived in force. Those from the federal safety agency called OSHA were most in evidence; they had
been present from the start, and had been largely ignored, but
were suddenly multiplying now and gaining the upper hand.
They wore bright safety vests and had helmets equipped with red
flashing lights. One afternoon, with about a dozen of them in
sight, their lights blinking in the hole, Pablo Lopez said to me,
"Look! The Martians have landed and they're communicating!" A
few days later one of them asked me to don safety glasses or

leave the excavation site, and I remember my surprise when I realized that he was serious. It felt sort of silly, like being required to wear sunblock in a combat zone, but the truth was that the battle was over, and the hole had become a tame place. Lopez's partner Andrew Pontecorvo explained it to me as a fact of life that he had observed before. He said, "The safer things get, the greater the restrictions." He was a realist. He shrugged.

It was a tough time for some of the victims' families, to whom it was increasingly clear that their missing would never be found. The firemen also were feeling that same frustration, and with less to do inside the hole, rather than reducing their forces they were concentrating with increasing zeal on raking through the debris that was laid out for secondary inspections in the southeast corner of the site. Relations between the tribes remained fractious. Among themselves, construction people accused the firemen of dragging things out for overtime pay, and the firemen, in turn, accused the construction men of profiteering. A gang of Port Authority cops attacked and severely beat a construction worker and a DDC fieldman who tried to come to his rescue, sending both men to the hospital. Once in early spring, when some New York City policemen refused to break off the recovery of one of their own dead to honor the remains of a fireman being carried up the bridge, a brawl broke out in the hole. The uniformed groups especially seemed sometimes to be clinging to their tragedies.

The larger mood at the site was based more on the recovery process itself than on the attack, and it was a complex mixture of nostalgia, satisfaction, and regret. People felt that this had been the peak experience of their lives, that they had "gone to the moon," as they said, and that they would never do anything to equal it again. Bill Cote told me about an encounter he had with Jim Miller, the big, blunt, blue-collar founder of a small company named Angel Aerial, normally in the business of producing theatrical rain (Victoria's Secret ads, for instance, or the fishmarket scene in *Godzilla*), who through sheer force of personality had busted into the dust-suppression role at the site. In the spring, with dust no longer such a problem, Cote required Miller to lower his prices or leave the job. Miller did not complain. Rather, he came up to thank Cote for the opportunity to stay. He said, "You don't understand what this means to me. I won't even bargain with you on it. I'll lower my price right now—what do you want it to be? Because I don't want to leave. I *can't* leave."

Of course, everyone eventually did have to leave, and it was that certainty coming into view that defined the final mood. Ken Holden was dealing with the consequences back at DDC headquarters in Queens, to which most of his emergency team had already returned. The people who were feeling listless and depressed were not necessarily those who had been most exposed to destruction and death on the pile but, rather, those who had worked in the classrooms of PS 89 and were faced now with returning to a workplace of fluorescent-lit cubicles and networked computers—an environment that, paradoxically, was all too much like that of the World Trade Center before the collapse.

Holden himself was riding high. During the last months of the Giuliani administration he had played a dangerous defense against the powerful and politically connected Bechtel Corporation, the San Francisco–based civil-engineering giant. Through mechanisms that were never made clear but presumably had to do with Republican contacts in Washington, Bechtel had gained backing within City Hall, and had repeatedly and aggressively attempted to insert itself into the Trade Center operation as an additional (and costly) management layer between the DDC and the construction companies. This was widely seen at the Trade

Center site as an intrusion so unnecessary that it could be understood only as the worst kind of opportunism. Holden fought back hard, using a variety of tactics, including indiscreetly wondering how the "sweetheart deal" headlines might read. And he won, though at considerable risk to his own career. For several months afterward, during which Bloomberg took office, Holden was on ice—managing the DDC and the Trade Center operation on a "transition" basis, and rather miserably looking for a job. By late winter, however, with a flurry of favorable reports mentioning his name in the press, things took a turn for the better and he was reappointed. Now again he was a happy man, swimming in the sea of New York politics, and with the Trade Center recovery to his credit, was full of new plans and ambitions for his beloved DDC.

Mike Burton was having a harder time letting go. One of his friends had warned him of the difficulty he now faced: as the publicly anointed Trade Center Czar, he would find it hard when someone said to his face, "Mike *who*?"—and that day would come soon. By the spring Bovis had taken over the day-to-day management of the operation, and though Burton was still in charge, he had fewer decisions to make; there were times now when he seemed cut off and alone. What was left of the DDC team had moved out of PS 89 and into temporary quarters high in the American Express building overlooking the site. I found Burton there one afternoon among the silent, empty cubicles, standing at a window as if he had nothing to do. I stood beside him, looking down at the hole, where the once entombed PATH train now stood in daylight. I'd been gone for a few days, and I asked him if anything had changed in my absence. He answered, "Oh, yeah, a lot!" But the truth was, nothing had. Things were slowing down.

Still, it was Burton's prerogative to see this job through. On that point even Ken Holden agreed. Of course Burton was never going to return to DDC headquarters. Neither he nor Holden could have stood it anymore. Holden tried to be gracious. When the Bloomberg administration called to inquire about Burton, Holden recommended him for higher public office—though perhaps with reservations. In the spring Bloomberg offered Burton a position as head of the city's Department of Buildings, considered an intractable bureaucracy in need of reform. Burton wisely turned it down. He talked to me a little dreamily about starting a company to respond to emergencies worldwide—he seemed to be searching for a way to relive the experience of the first few months. Many people at the site had that same wistful thought. In the end, though, Burton had his family and ambitions to look after. With Holden's blessing, he stayed on for several weeks beyond the closing ceremony, and in the summer finally quit the DDC to become a senior vice-president for the same large construction company he had worked for before joining the city government. It was the standard revolving-door arrangement. He would be based in New York and responsible for regional operations, along with the company's growing "homeland security" business. He would also make a lot of money. And that, too, had been predictable for months.

Sam Melisi ended the job in some ways worse off than he had begun: as the emotions had continued to intensify at the site, he had come under attack by some of the more extreme firemen and widows, who felt that he had turned on them—that the very act of listening to their opponents was a form of betrayal. Unwilling to see the search come to an end, some of them began hunting for traitors in their midst—and as often happens, they turned on the mediator. Melisi was tainted by association. Of course they completely misjudged the man. He was an empathetic person, it was true, and because of his own background in construction he understood the mentality of the unbuilders at the site, but his only real allegiances had always been to the firemen and the families of the dead. Though he refused to complain about the accusations now, they must have been difficult for him to bear. No doubt, however, it was for more physical reasons—call it absolute exhaustion—that in April he suffered a mild heart attack. He was put on medical leave, and had to withdraw from the site. I went to see him for dinner one evening at his house on Staten Island. He told me he wanted nothing more now than to get a clean bill of health and return to being a front-line fireman. His plan was to join one of the heavily hit rescue squads in Brooklyn, even though this would mean a cut in pay, because he felt he could be of use there. But the truth was that even such humble goals were now in doubt. After dinner the conversation drifted to the meaning of it all, and the subject of history came up. He said he hadn't cared about it before, but cared about it now. He said he sometimes worried about an apocalyptic future. The conversation might then have become too lofty for either of our tastes, were it not for the children at the table, Melisi's young son and daughter, who were bored by all the talk. They wanted their chocolate cake. The boy showed us a card trick. Melisi forgot about history, and simply got on with living.

The next morning I went to the pier of a company called Metal Management, on the shores of Newark Bay, to watch the Trade Center being sent away. It was steel that was going—a load of the heavy perimeter columns from both of the Twin Towers. Cutting crews were torching them into three-foot sections, sized to fit into the charge boxes of steel-mill furnaces overseas. The sections they cut lay in hills on the pier, with rust already setting in. An old diesel loader rumbled back and forth, moving them into a huge wedge-shaped pan, which was then lifted by crane and tilted to dump the steel into the cargo holds of an aging 500-foot ship that lay alongside the pier. With the exception of a few samples that had been held for the engineering investigations, this was the fate of almost all of the World Trade Center's 200,000 tons of structural steel, the columns and girders that had given the buildings their strength and then, wounded, allowed them to fall. It was exceptional steel, some of the purest these cutters had ever seen, but too expensive for American mills to reuse, because of the high costs of recycling in the United States. Costs were lower elsewhere, of course, and labor and environmental rules were more relaxed. For better or worse, the scrap-metal market is famously global.

The ship was the *Osman Mete*, which hailed from Istanbul with a Turkish crew, and not by chance. Turkish shippers were said to be less sensitive than others to the damage that could be done to their vessels by such ultra-heavy scraps. At the very least, they were accustomed to such cargoes, since Turkish mills had been among the first to acquire and melt down some of the Trade Center's structure. With this load now, however, the *Osman Mete* was heading in the opposite direction, forty days through the Panama Canal and on to China, which along with India had turned out to be the principal destination for the Twin Towers. I scrambled up the pilot ladder and onto the deck, and for several hours just stood there and watched. The ship was rusty and badly maintained. Its deck machinery looked like it had been broken for years. The crew was filthy, and obviously indifferent to the meaning of this load. The hatches lay wide open. Seen from above, the holds were cavernous and badly battered.

Every few minutes the crane brought the loaded pan overhead and emptied it. The steel tumbled down with a roar, and sent shudders through the ship. The dust that rose had the old sweet smell of the pile. After it cleared, the steel became visible again, lying haphazardly where it had fallen, already in foreign hands, and destined for furnaces on the far side of the globe. It was a strangely appropriate fate for these buildings. Unmade or re-made, whether as appliances or cars or simple rebar, they would eventually find their way into every corner of the world.